High Frequency Communication and Sensing

Devices, Circuits, and Systems

Series Editor
Krzysztof Iniewski
CMOS Emerging Technologies Research Inc.,
Vancouver, British Columbia, Canada

PUBLISHED TITLES:

Atomic Nanoscale Technology in the Nuclear Industry
Taeho Woo

Biological and Medical Sensor Technologies
Krzysztof Iniewski

Building Sensor Networks: From Design to Applications
Ioanis Nikolaidis and Krzysztof Iniewski

Circuits at the Nanoscale: Communications, Imaging, and Sensing
Krzysztof Iniewski

Design of 3D Integrated Circuits and Systems
Rohit Sharma

Electrical Solitons: Theory, Design, and Applications
David Ricketts and Donhee Ham

Electronics for Radiation Detection
Krzysztof Iniewski

**Embedded and Networking Systems:
Design, Software, and Implementation**
Gul N. Khan and Krzysztof Iniewski

Energy Harvesting with Functional Materials and Microsystems
Madhu Bhaskaran, Sharath Sriram, and Krzysztof Iniewski

**Graphene, Carbon Nanotubes, and Nanostuctures:
Techniques and Applications**
James E. Morris and Krzysztof Iniewski

High-Speed Devices and Circuits with THz Applications
Jung Han Choi

High-Speed Photonics Interconnects
Lukas Chrostowski and Krzysztof Iniewski

**High Frequency Communication and Sensing:
Traveling-Wave Techniques**

Ahmet Tekin and Ahmed Emira

Integrated Microsystems: Electronics, Photonics, and Biotechnology
Krzysztof Iniewski

PUBLISHED TITLES:

Integrated Power Devices and TCAD Simulation
Yue Fu, Zhanming Li, Wai Tung Ng, and Johnny K.O. Sin

Internet Networks: Wired, Wireless, and Optical Technologies
Krzysztof Iniewski

Labs-on-Chip: Physics, Design and Technology
Eugenio Iannone

Low Power Emerging Wireless Technologies
Reza Mahmoudi and Krzysztof Iniewski

Medical Imaging: Technology and Applications
Troy Farncombe and Krzysztof Iniewski

Metallic Spintronic Devices
Xiaobin Wang

MEMS: Fundamental Technology and Applications
Vikas Choudhary and Krzysztof Iniewski

Micro and Nanoelectronics: Emerging Device Challenges and Solutions
Tomasz Brozek

Microfluidics and Nanotechnology: Biosensing to the Single Molecule Limit
Eric Lagally

MIMO Power Line Communications: Narrow and Broadband Standards, EMC, and Advanced Processing
Lars Torsten Berger, Andreas Schwager, Pascal Pagani, and Daniel Schneider

Mobile Point-of-Care Monitors and Diagnostic Device Design
Walter Karlen

Nano-Semiconductors: Devices and Technology
Krzysztof Iniewski

Nanoelectronic Device Applications Handbook
James E. Morris and Krzysztof Iniewski

Nanopatterning and Nanoscale Devices for Biological Applications
Šeila Selimović

Nanoplasmonics: Advanced Device Applications
James W. M. Chon and Krzysztof Iniewski

Nanoscale Semiconductor Memories: Technology and Applications
Santosh K. Kurinec and Krzysztof Iniewski

Novel Advances in Microsystems Technologies and Their Applications
Laurent A. Francis and Krzysztof Iniewski

Optical, Acoustic, Magnetic, and Mechanical Sensor Technologies
Krzysztof Iniewski

Radiation Effects in Semiconductors
Krzysztof Iniewski

PUBLISHED TITLES:

Semiconductor Radiation Detection Systems
Krzysztof Iniewski

Smart Grids: Clouds, Communications, Open Source, and Automation
David Bakken

Smart Sensors for Industrial Applications
Krzysztof Iniewski

Technologies for Smart Sensors and Sensor Fusion
Kevin Yallup and Krzysztof Iniewski

Telecommunication Networks
Eugenio Iannone

Testing for Small-Delay Defects in Nanoscale CMOS Integrated Circuits
Sandeep K. Goel and Krishnendu Chakrabarty

VLSI: Circuits for Emerging Applications
Tomasz Wojcicki

Wireless Technologies: Circuits, Systems, and Devices
Krzysztof Iniewski

FORTHCOMING TITLES:

Analog Electronics for Radiation Detection
Renato Turchetta

**Cell and Material Interface: Advances in Tissue Engineering,
Biosensor, Implant, and Imaging Technologies**
Nihal Engin Vrana

Circuits and Systems for Security and Privacy
Farhana Sheikh and Leonel Sousa

CMOS: Front-End Electronics for Radiation Sensors
Angelo Rivetti

CMOS Time-Mode Circuits and Systems: Fundamentals and Applications
Fei Yuan

**Electrostatic Discharge Protection of Semiconductor Devices
and Integrated Circuits**
Juin J. Liou

Gallium Nitride (GaN): Physics, Devices, and Technology
Farid Medjdoub and Krzysztof Iniewski

Implantable Wireless Medical Devices: Design and Applications
Pietro Salvo

Laser-Based Optical Detection of Explosives
Paul M. Pellegrino, Ellen L. Holthoff, and Mikella E. Farrell

FORTHCOMING TITLES:

Mixed-Signal Circuits
Thomas Noulis and Mani Soma

MRI: Physics, Image Reconstruction, and Analysis
Angshul Majumdar and Rabab Ward

Multisensor Data Fusion: From Algorithm and Architecture Design to Applications
Hassen Fourati

Nanoelectronics: Devices, Circuits, and Systems
Nikos Konofaos

Nanomaterials: A Guide to Fabrication and Applications
Gordon Harling, Krzysztof Iniewski, and Sivashankar Krishnamoorthy

Optical Fibre Sensors: Advanced Techniques and Applications
Ginu Rajan

Organic Solar Cells: Materials, Devices, Interfaces, and Modeling
Qiquan Qiao and Krzysztof Iniewski

Power Management Integrated Circuits and Technologies
Mona M. Hella and Patrick Mercier

Radiation Detectors for Medical Imaging
Jan S. Iwanczyk and Polad M. Shikhaliev

Radio Frequency Integrated Circuit Design
Sebastian Magierowski

Reconfigurable Logic: Architecture, Tools, and Applications
Pierre-Emmanuel Gaillardon

Soft Errors: From Particles to Circuits
Jean-Luc Autran and Daniela Munteanu

Solid-State Radiation Detectors: Technology and Applications
Salah Awadalla

Wireless Transceiver Circuits: System Perspectives and Design Aspects
Woogeun Rhee and Krzysztof Iniewski

High Frequency Communication and Sensing

TRAVELING-WAVE TECHNIQUES

Ahmet Tekin
Waveworks, Inc.
Sunnyvale, California, USA

Ahmed Emira
Cairo University
Giza, Egypt

CRC Press
Taylor & Francis Group
Boca Raton London New York

CRC Press is an imprint of the
Taylor & Francis Group, an **informa** business

CRC Press
Taylor & Francis Group
6000 Broken Sound Parkway NW, Suite 300
Boca Raton, FL 33487-2742

First issued in paperback 2017

© 2015 by Taylor & Francis Group, LLC
CRC Press is an imprint of Taylor & Francis Group, an Informa business

No claim to original U.S. Government works

ISBN-13: 978-1-4822-0711-8 (hbk)
ISBN-13: 978-1-138-89370-2 (pbk)

Library of Congress Cataloging-in-Publication Data

Tekin, Ahmet.
　High frequency communication and sensing : traveling-wave techniques / authors, Ahmet Tekin and Ahmed Emira.
　　　pages cm -- (Devices, circuits, and systems ; 35)
　Includes bibliographical references and index.
　ISBN 978-1-4822-0711-8 (hardback)
　1. Microwave communication systems. 2. Microwave circuits. 3. Traveling wave antennas. 4. Sensor networks. I. Emira, Ahmed. II. Title.

TK5103.4833.T45 2014
621.381'3--dc23　　　　　　　　　　　　　　　　　　　　　　　　　　　　2014019857

Visit the Taylor & Francis Web site at
http://www.taylorandfrancis.com

and the CRC Press Web site at
http://www.crcpress.com

Contents

Preface xi
About the Authors xiii

1. **What This Book Is About** 1

2. **Lumped vs. Distributed Elements** 7
   2.1   Infinite Transmission Line 7
   2.2   Dispersionless Transmission Line 9
   2.3   Lossless Transmission Line 11
   2.4   Shorted Transmission Line 13
   2.5   Voltage Standing Wave Ratio (VSWR) 14
   2.6   Smith Chart 16
   Bibliography 21

3. **Trigger Mode Distributed Wave Oscillator** 23
   3.1   Commonly Used Wave Oscillator Topologies 26
   3.2   Rotary Traveling Wave Oscillator (RTWO) 27
   3.3   Standing Wave Oscillator (SWO) 31
   3.4   TMDWO Topology 34
   3.5   Phase Noise in Traveling Wave Oscillators 38
          3.5.1   Phase Noise in SWO 38
                   3.5.1.1   Noise of Transmission Line 41
                   3.5.1.2   Tail Transistor Noise 42
                   3.5.1.3   Cross-Coupled Pair Noise 43
          3.5.2   Phase Noise in RTWO 46
          3.5.3   Phase Noise in Differential Wave Oscillator (DWO) 51
   3.6   Experimental Results 53
   3.7   Conclusion 55
   Bibliography 55

4. **Force Mode Distributed Wave Oscillator** 57
   4.1   Force Mode Distributed Wave Oscillation Mechanisms 57
   4.2   Single-Ended Force Mode Structures 59
   4.3   Conclusion 68
   Bibliography 69

5. **Wave-Based RF Circuit Techniques** 71
    5.1   Pumped Distributed Wave Oscillators 72
    5.2   Traveling Wave Phased-Array Transceiver 74
    5.3   Conclusion 81
    Bibliography 81

6. **THz Signal Generation and Sensing Techniques** 83
    6.1   Frequency Multiplication Techniques 84
    6.2   Traveling Wave Reflectometers 88
    6.3   Wafer-Level THz Sensing Method 95
    6.4   Conclusion 96
    Bibliography 96

7. **Traveling Wave-Based High-Speed Data Conversion Circuits** 99
    7.1   Traveling-Wave Noise Shaping Modulator 107
    7.2   A High-Speed Phase Interleaving Topology 109
    7.3   A Traveling Wave Multiphase DAC 115
    7.4   Traveling Wave Phased-Array DAC Transmitter 117
    7.5   Conclusion 119
    Bibliography 119

8. **Traveling Wave High-Speed Serial Link Design for Fiber
    and Backplane** 121
    8.1   Traveling Wave-Based Multiphase Rx-Tx Front End 126
    8.2   A Full-Rate Phase-Interpolating Topology 128
    8.3   An ADC-Based DSP Link Front End 130
    8.4   Conclusion 131
    Bibliography 131

Index 133

Preface

Telecommunication and medical applications have become two main driving forces for emerging semiconductor technologies. Computational power slowly shifting into the cloud servers has made communication speed a limitation in personal computing. Hence, demand for speed from backhaul serial optical wireline networks has experienced a surge. The existing optical networks with 10 Gb/s per lane link rates has been targeted to boost the system with a 40 Gb/s and even 100 Gb/s per lane data rate over the existing fiber optics infrastructure. This, however, has pushed the burden to the interface integrated circuits, demanding more speed with less noise margins. The spread of many WiFi and cellular small cell standards on the wireless side has also boosted the effort for more wireless channel capacity per available spectrum. One remedy in the search for more throughputs on the wireless side is to allocate some new, very high frequency bands, mainly around 60 GHz range, as an unlicensed free spectrum. Although a relatively large chunk of spectrum has become available for local wireless networks, design of high-speed integrated circuits at such frequencies remains a challenge. This book introduces some new traveling wave circuit techniques to boost the performance of high-speed circuits in standard low-cost production technologies like the complementary metal oxide semiconductor (CMOS). While the book serves as a valuable resource for experienced analog/radio frequency (RF) circuit designers, it also provides much information for undergraduate-level microelectronics researchers.

The main theme of this book involves utilizing multiple phases of traveling waves as a fine high-speed timing reference rather than carrying high-speed signals across a microchip. It is a significantly challenging task to generate and distribute high-speed clocks. Multiphase low-speed clocks with sharp transition are proposed to be a better option to accommodate the desired timing resolution. The authors believe that the proposed techniques will provide new horizons in the quest for more speed and performance.

Chapter 2 describes high-speed signaling basics, such as transmission lines, distributed signaling, impedance matching, and other common practical RF background material. This chapter introduces entry-level readers to the chapters that follow.

In Chapter 3, a dual-loop coupled traveling wave oscillator topology, the trigger mode distributed wave oscillator, is described in detail as a high-frequency multiphase signal source. Thanks to the proposed trigger-mechanisms, two independent transmission line oscillators can be cross-coupled to form a single differential oscillator. Multiple oscillation phases become readily available along these symmetric oscillation tracks. These clock phases are to be used in much of the work presented in the following chapters. The advantages and disadvantages of such a traveling wave oscillator are described in detail.

Chapter 4 introduces a force-based starter mechanism for dual-loop, even-symmetry, multiphase traveling wave oscillators. A single-loop version of this oscillator is presented as a force mode distributed wave antenna (FMDWA). Moreover, coupling this single-ended structure with secondary pickup coils results in various integrated microwave transformer configurations. A phased-array transmitter system is also presented as one of the applications of the mentioned traveling wave structure.

In Chapter 5, even higher-frequency, passive inductive, or quarter-wavelength-based pumped distributed wave oscillators (PDWOs) are described in detail. As in the case of other wave oscillators, these techniques also provide multiple high-frequency oscillation phases. Phased-array transceiver architectures and front-end circuits are illustrated in detail along with the distributed oscillator topologies.

Chapter 6 is devoted to THz sensing. A unique method of traveling wave frequency multiplication and power combining is illustrated. The mentioned THz signal source is then employed in a reflection-based integrated sensing transceiver device that forms a unit pixel element for a low-cost, nonradioactive imaging device. Creating a large array of these elements on a silicon wafer will result in a complete THz image sensor.

Chapter 7 discusses various data converter topologies such as digital-to-analog converters (DACs), analog-to-digital converters (ADCs), and GHz-bandwidth sigma-delta modulators using the integrated multiphase clock sources described in previous chapters. Low-noise clock phases this time provide opportunity for high-accuracy interleaved time intervals for the mentioned sampling systems.

Another set of serial link sampling systems is discussed in Chapter 8. Once more, many symmetric phases of the traveling wave oscillators find a special place in circuit blocks employed in these systems. Phase rotators and interpolators, phase shifters, phase-locked loops (PLLs), and delay-locked loops (DLLs) are some of the critical circuits that are discussed.

About theAuthors

Ahmet Tekin received his EE PhD degree from the University of California Santa Cruz, California, EE MS degree from North Carolina A&T State University, Greensboro, North Carolina, and EE BS degree from Bogazici University, Istanbul, Turkey, in 2008, 2004, and 2002, respectively. During his MS studies at NCA&T State University, he worked on a NASA transceiver project designing very low power radiation-hard SOI complementary metal oxide semiconductor (CMOS) circuits. While obtaining his PhD, he designed a very low noise analog radio baseband with noise-shaping circuit techniques. He has worked for many innovative companies: Multigig, Inc. on the technical staff; Newport Media as a senior analog design engineer; Aydeekay LLC as a senior mixed-signal design engineer; Broadcom Corporation as a senior staff scientist, Semtech Corporation as a principle design engineer; and Nuvoton Technology Corporation as an analog design manager. He was co-founder of Waveworks, Inc., focusing on novel traveling wave-based high-frequency communication circuits and biochips.

Ahmed Emira received his BSc and MSc degrees in electronics and communications from Cairo University, Giza, Egypt, in 1997 and 1999, respectively. In December 2003, he received his PhD degree in electrical engineering at Texas A&M University, College Station, Texas. From 2001 to 2002 he was with Motorola, Austin, Texas, where he worked as a radio frequency integrated circuit (RFIC) design engineer. Following his PhD, he worked as an RFIC design engineer in the wireless division of Silicon Laboratories, Austin, Texas, from 2003 to 2006. Then he worked as a senior RFIC design engineer and a leader for the power management team in Newport Media, Inc., Lake Forest, California, from 2006 to 2008. He joined Cairo University as an assistant professor in the electronics and communications department in 2008. Dr. Emira is currently the manager of the RFIC group at Atmel Inc., Egypt Design Center, and is also an associate professor at Cairo University. He has more than 40 journal and conference publications, 5 U.S. patents, and several pending. His current interests include low-power mixed-signal circuits for portable devices/energy harvesting systems, mm-wave and RF circuits, MEMS interface electronics, and wireless communication system architectures.

1

What This Book Is About

Communications, one form or other, has been one of the most valuable tools for man. The speed of developments in the last two decades, though, was unmatched. The most cited inventions of the century have been wireless cellular communications and the Internet. It could only be science fiction for this living generation 30 years ago that one would be able to experience a live conversation with a family or friend on the other side of the world in the near future. Although we can enjoy an amazing amount of communication experience inexpensively through a small handheld device today, the cost for these technologies to reach this point was not low. Generations of engineers have worked years to come upon every incremental improvement to make this experience smoother and more enjoyable. From transatlantic under-ocean links and satellites to cellular and wireless local area network (LAN) technologies, the incredible amount of infrastructure web around the world did not stop humans from desiring more. The increasing demand for wireless communication has made the frequency spectrum one of the most valuable commodities in recent human history. Hence, the technologies that can help in better utilization of this valuable resource deserve valuable recognition as well. Figure 1.1 shows the crowding in today's wireless spectrum. Most of the spectrum is allocated with very fine spacing between various standards, and the channel widths are all limited. Heavy traffic around the ultra-high frequencies (UHFs) does not allow much possibility for wider channels. Even the communication in the existing channels can be hijacked by nearby interferers in adjacent frequency bands. The propagation characteristics at these frequencies have proven to be optimal for long-range communication; hence, most 2G, 3G, and 4G cellular networks are centered at around these frequencies. However, as the demand for more bandwidth grows, the microwave frequency bands come into play as resources to be tapped despite the hardware design challenges and cost limitations. The 23 and 38 GHz microwave bands are two strong candidates for the emerging 5G femtocell or picocell type of cellular networks that are under review across the globe. Moreover, the 7 to 10 GHz industrial, scientific, and medical (ISM) band around 60 GHz has long been targeted by many technology companies since its release by the Federal Communications Commission (FCC) and its counterparts around the world; however, no widespread deployment of hardware is recorded yet due to the design challenges at such high frequencies. High oxygen absorption and poor radiation characteristics are only a few to mention.

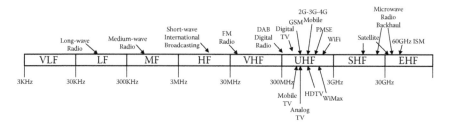

FIGURE 1.1
Today's wireless communication spectrum.

Wi-Fi LAN devices constitute a significant portion of wireless devices around us. The standard is called 802.11 after the name of the group formed to oversee its development. First, 802.11 only supported a maximum network bandwidth of 2 Mbps. For this reason, the initial version of 802.11 wireless products are no longer available. The Institute of Electrical and Electronics Engineers (IEEE) expanded on the original 802.11 standard in July 1999, creating the 802.11b specification. 802.11b supports bandwidths up to 11 Mbps. 802.11b uses the same ISM band at 2.4 GHz as the original 802.11 standard. 802.11b devices can incur interference from microwave ovens, cordless phones, Bluetooth, and other appliances using the same 2.4 GHz range. While 802.11b was in development, IEEE created a second extension to the original 802.11 standard called 802.11a. 802.11b gained much more popularity than 802.11a due to its lower cost. 802.11a was used on business networks, whereas 802.11b better served the home market. 802.11a supports bandwidths up to 54 Mbps and signals in a regulated frequency spectrum of around 5 GHz. This higher frequency compared to 802.11b shortens the range of 802.11a networks. The higher frequency also means 802.11a signals have more difficulty penetrating walls and other obstructions. Because 802.11a and 802.11b utilize different frequencies, the two technologies are incompatible with each other. In 2003, a new wireless (WLAN) standard, called 802.11g, hit the market. 802.11g combines the best of both 802.11a and 802.11b. 802.11g supports bandwidths up to 54 Mbps, similar to 802.11a, but it uses the 2.4 GHz frequency for long-range transmission. 802.11g is backwards compatible with 802.11b, meaning that 802.11g access points will work with 802.11b wireless network adapters.

802.11n came along to improve on 802.11g in the amount of bandwidth supported by utilizing multiple-input multiple-output (MIMO) technology. Standards groups ratified 802.11n in 2009 with specifications providing for up to 300 Mbps of network bandwidth. 802.11n also offers somewhat better range over earlier Wi-Fi standards due to its increased signal intensity, and it is backwards compatible with 802.11b/g. One of the latest standards, 802.11ac utilizes dual-band wireless technology, supporting simultaneous connections on both the 2.4 and 5 GHz Wi-Fi bands. 802.11ac offers backward compatibility to 802.11b/g/n and bandwidths rated up to 1300 Mbps on the 5 GHz band, plus up to 450 Mbps on 2.4 GHz. In addition to these four Wi-Fi standards, other related wireless network technologies have to share the same 2.4 GHz ISM band. Bluetooth, for example, uses 2.4 GHz. Bluetooth supports a very short range of up to 10 m and relatively low bandwidth of 1 Mbps. It is used in low-power network devices such as medical devices, wireless keyboards and mouses, and handhelds. Other than 5 and 2.4 GHz ISM bands, many other bands are used for machine-to-machine wireless connectivity solutions. Some of the most used ISM bands around the world are 315 and 915 MHz in the United States; 187, 230, 433, and 868 MHz in Europe; 426, 429, 449, 950, and 1200 MHz in Japan; and 223, 230, 315, and 433 MHz in China. These devices are generally deployed in cost-sensitive consumer products such as garage door openers, remote keyless entry, toys, and remote controllers. Mobile cellular networks also consume a significant amount of spectral resources due mainly to the number of active users at a given time. Initial forms of cellular networks were used to deliver voice-only data through analog modulation; hence, the sound quality was poor and the speed of transfer was only at 9.6 Kbps. With the advance of 2G networks, the transmission quality had improved by introducing the concept of digital modulation. 2.5G is a transition between 2G and 3G. In 2.5G, the most popular services, like Short Message Service (SMS), General Packet Radio Service (GPRS), Enhanced Data for GSM (Global System for Mobile Communication) Evolution (EDGE), and more, had been introduced. Shortly after the introduction of 2.5G systems, the 3G generation of mobile telecommunication standards had emerged. It allows simultaneous use of speech and data services and offers data rates of up to 2 Mbps, which provide services like video calls, mobile TV, mobile Internet, and downloading. There are many technologies that fall under 3G, like Wideband Code Division Multiple Access (WCDMA) and High Speed Package Access (HSPA). In telecommunications, 4G is the fourth generation of cellular wireless standards. It is a successor to the 3G and 2G families of standards. In 2008, the ITU-R (International Telecommunication Union Radiocommunication Sector) organization specified the IMT-Advanced (International Mobile Telecommunications Advanced) requirements for 4G standards, setting data transfer requirements for 4G service at 100 Mbit/s for high-mobility communication, such as from trains and cars, and 1 Gbit/s for low-mobility communication stationary users. A 4G system is expected to provide a comprehensive and secure all-(Internet Protocol) IP-based mobile broadband solution to laptop computers, wireless modems, smartphones, and other mobile devices. New services such as ultra-broadband Internet access, IP telephony, gaming services, and streamed multimedia may be provided to users. Early 4G technologies such as mobile WiMAX and Long-term Evolution (LTE) have been around for a while. The current versions of these technologies did not fulfill the original ITU-R requirements of data rates approximately up to 1 Gbit/s for 4G systems. Another wireless communication field is GPS. All satellites broadcast at the same two frequencies, 1.57542 GHz (L1 signal) and 1.2276 GHz

(L2 signal). The satellite network uses a code division multiple access (CDMA) spread-spectrum technique where the low-bitrate message data are encoded with a high-rate pseudorandom number (PRN) sequence that is different for each satellite. The receiver must be aware of the PRN codes for each satellite to reconstruct the actual message data. The L1 carrier is modulated by both the C/A and P codes, while the L2 carrier is only modulated by the P code. The P code can be encrypted as a so-called P(Y) code that is only available to military equipment with a proper decryption key. Both the C/A and P(Y) codes impart the precise time of day to the user. The L3 signal at a frequency of 1.38105 GHz is used to transmit data from the satellites to ground stations. These data are used by the U.S. Nuclear Detonation Detection System (USNDS) to detect, locate, and report nuclear detonations (NUDETs) in the earth's atmosphere and near space. One usage is the enforcement of nuclear test ban treaties. The L4 band at 1.379913 GHz is being studied for additional ionospheric correction. The L5 frequency band at 1.17645 GHz was added in the process of GPS modernization. This frequency falls into an internationally protected range for aeronautical navigation, promising little or no interference under all circumstances. The L5 consists of two carrier components that are in phase quadrature with each other. Each carrier component is bi-phase shift key (BPSK) modulated by a separate bit train.

High operating frequencies, and hence the required hardware speeds, have become a challenge not only for wireless networks but also for wireline networks. Ethernet and optical communication networks constantly seek higher and higher speeds. It should be noted that in addition to the wireline Internet traffic, the mentioned wireless cellular traffic has to eventually flow through these networks at the backhauls. Newly emerging connectivity solutions such as cloud servers, Internet-based TV, social media servers, video conferencing technologies, and online gaming have put even more pressure on these wireline Ethernet/optical networks. Recently, for example, the IEEE 802.3 400Gb/s Ethernet Study Group was formed to explore development of a 400Gb/s Ethernet standard to address the demand for more bandwidth. Although the technology scaling has helped to track the demand well in recent decades, Moore's law has started to break apart from its original trend due to the physical limitations of lithography. Hence, more is needed at the design side to catch up with the projected demand for the mentioned communication technologies. This book introduces many unique, high-performance traveling wave-based circuits, such as multi phase oscillators, transceivers, interleaved analog-to-digital converters (ADCs), and interleaved digital-to-analog converters (DACs) for future communication electronics. Special types of traveling wave oscillators presented in the chapters ahead can provide very high frequency signals in a low-cost standard complementary metal oxide semiconductor (CMOS) process.

In addition to communications, high-frequency signals have found a new role in the emerging field of THz sensing. Figure 1.2 shows the electromagnetic spectrum. There is an interesting electromagnetic band in the range from 100

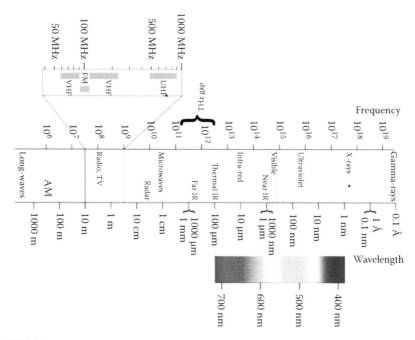

FIGURE 1.2
Electromagnetic spectrum.

up to 10 GHz that contains unique frequencies serving as specific markers for many medical and scientific applications. This band is nowadays called 'THz gap' since neither electronics nor optics can handle it well. High-speed low-cost electronics still struggle at sub-THz speeds, whereas optics most commonly requires multiple-source frequency translation to generate THz signals. Significant research effort is underway to close the THz gap with various techniques, but no low-cost efficient way is available yet.

Some microwave sensing circuits were introduced as well to be the basis for future sensing applications such as THz cancer detection and medical imaging, chemical and explosive detection, space research, metal detectors, and security systems. The enabling element in these systems is the THz signal source. Traveling wave-based THz signal generation and utilization circuits were introduced.

2

Lumped vs. Distributed Elements

Lumped circuit models, such as resistors, capacitors, and inductors, are used to evaluate the response of a circuit relative to low frequency. At sufficiently high frequencies, the lumped model of an element is no longer an accurate representation of its behavior. More specifically, this occurs when the operation frequency is comparable with the traveling speed of the voltage and current waves across the element boundaries. In other words, when the wavelength of the traveling wave becomes comparable with the element dimensions, we can no longer assume that the voltage or current waveforms are constant across the element dimension. In this case, we must consider modeling the element using distributed circuit model. The distinction between lumped and distributed elements can be best understood by studying the traveling wave in a transmission line [Collin (1966), G. D. Vendelin and Pavio (1990), and Pozar (1997)].

2.1 Infinite Transmission Line

Suppose we have an infinitely long transmission line, which is driven with a sinusoidal voltage source as shown in Figure 2.1. The transmission line can be modeled as an infinite number of cascaded sections of infinitesimal length dx. Each section can be modeled with a series impedance $Z_s dx$ and a shunt admittance $Y_p dx$, as illustrated in Figure 2.2, where Z_s and Y_p are the series impedance and shunt admittance per unit length. The input impedance of this section can be written as

$$Z_{in}(x) = \frac{1}{Y_p dx + 1/(Z_{in}(x+dx) + Z_s dx)} \tag{2.1}$$

$$Z_0 Y_p dx + \frac{Z_0}{Z_0 + Z_s dx} = 1 \tag{2.2}$$

Using the fact that $Z_s dx \ll Z_0$, the second term in the left-hand side of Equation (2.2) is expanded with the first two terms of the Taylor series as

$$Z_0 Y_p dx + 1 - \frac{Z_s dx}{Z_0} = 1 \tag{2.3}$$

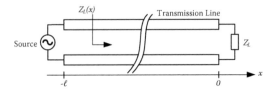

FIGURE 2.1
Transmission line with infinite length.

which can be simplified to

$$Z_0 = \sqrt{\frac{Z_s}{Y_p}} \tag{2.4}$$

This is called the characteristic impedance of the transmission line. It is interesting to note that the input impedance of an infinitely long transmission line is actually a finite value. It is also important to calculate the output voltage at distance $x + dx$ with respect to the voltage at x. By performing Kirchhoff's current laws (KCL) and Kirchhoff's voltage laws (KVL) in Figure 2.2, we obtain

$$V(x) = V(x + dx) + I(x + dx)Z_s dx \tag{2.5}$$

$$I(x) = I(x + dx) + V(x)Y_p dx \tag{2.6}$$

From Equations (2.5) and (2.6), we obtain the following first-order differential equations:

$$\frac{dV(x)}{dx} = -Z_s I(x + dx) \approx -Z_s I(x) \tag{2.7}$$

$$\frac{dI(x)}{dx} = -Y_p V(x) \tag{2.8}$$

Therefore, by combining Equations (2.7) and (2.8), we obtain the following second-order differential equation:

$$\frac{d^2 V(x)}{dx^2} = Z_s Y_p V(x) = \gamma^2 V(x) \tag{2.9}$$

$$\gamma = \sqrt{Z_s Y_p} \tag{2.10}$$

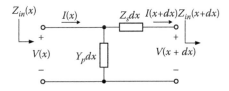

FIGURE 2.2
Circuit model of an infinitesimal section of the transmission line.

where γ is the propagation constant that represents the change in the wave amplitude and phase per unit length. The general solution for the above second-order differential equation is

$$V(x) = V_f e^{-\gamma x} + V_r e^{\gamma x} \tag{2.11}$$

By substituting from Equation (2.11) into Equation (2.7), we get the general current equation

$$I(x) = \frac{-1}{Z_s} \frac{dV(x)}{dx} = \frac{\gamma}{Z_s} \left(V_f e^{-\gamma x} - V_r e^{\gamma x} \right) = \frac{1}{Z_0} \left(V_f e^{-\gamma x} - V_r e^{\gamma x} \right) \tag{2.12}$$

The above solution is composed of two traveling waves; the first term represents a forward wave traveling toward the load (in the x-direction), while the second term represents the reverse wave traveling back to the source (opposite to the x-direction). The propagation constant is generally a complex term:

$$\gamma = \alpha + j\beta \tag{2.13}$$

where α is the attenuation constant that denotes the amplitude attenuation per unit length in the direction of the traveling wave. Specifically, the reverse wave amplitude drops as it travels in the x-direction, while the forward wave amplitude drops in the $-x$-direction. β is the phase constant that denotes the phase change per unit length of the transmission line. For a lossless transmission line, with $\alpha = 0$, the wave undergoes only a phase change as it travels through the transmission line. For the general case where $Z_s = j\omega L + R$ and $Y_p = j\omega C + G$, the propagation constant is expressed as

$$\gamma = \sqrt{(j\omega L + R)(j\omega C + G)} \tag{2.14}$$

And the characteristic impedance is expressed as

$$Z_0 = \sqrt{\frac{j\omega L + R}{j\omega C + G}} \tag{2.15}$$

2.2 Dispersionless Transmission Line

At sufficiently higher frequencies for a weakly lossy transmission line, we can assume that $R \ll \omega L$ and $G \ll \omega C$, and therefore we can approximate the propagation constant as

$$\gamma \approx j\omega\sqrt{LC}\left(1 + \frac{R}{j2\omega L}\right)\left(1 + \frac{G}{j2\omega C}\right) \approx j\omega\sqrt{LC} + \frac{R}{2}\sqrt{\frac{C}{L}} + \frac{G}{2}\sqrt{\frac{L}{C}} = \alpha + j\beta \tag{2.16}$$

Therefore, we can write

$$\alpha = \frac{R}{2}\sqrt{\frac{C}{L}} + \frac{G}{2}\sqrt{\frac{L}{C}} \tag{2.17}$$

$$\beta = \omega\sqrt{LC} \tag{2.18}$$

For a forward wave:

$$V(x) = V(0)e^{j\omega t}e^{-\alpha x}e^{-j\beta x} = V(0)e^{j(\omega t - \beta x)}e^{-\alpha x} = V(0)e^{j\theta(x,t)}e^{-\alpha x} \tag{2.19}$$

where $\theta(x, t)$ is total phase. To determine the wave propagation velocity along the transmission line, we can track a constant phase point and check how fast it travels.

$$\frac{d\theta}{dt} = \omega - \beta\frac{dx}{dt} = 0 \tag{2.20}$$

This constant phase point propagates in the x-direction at a velocity v_p, which is expressed as

$$v_p = \frac{dx}{dt} = \frac{\omega}{\beta} \approx \frac{1}{\sqrt{LC}} \tag{2.21}$$

Since the propagation velocity and the attenuation constant are independent of frequency, this transmission line is dispersionless. The general condition for dispersionless transmission can be obtained by forcing γ to be in the following form:

$$\gamma = j\frac{\omega}{v_p} + \alpha = \sqrt{(j\omega L + R)(j\omega C + G)} \tag{2.22}$$

where v_p and α in the above equation are independent of frequency. By squaring both sides of the equation:

$$\left(j\frac{\omega}{v_p} + \alpha\right)^2 = (j\omega L + R)(j\omega C + G) \tag{2.23}$$

$$2j\alpha\frac{\omega}{v_p} + \alpha^2 - \left(\frac{\omega}{v_p}\right)^2 = j\omega(LG + CR) + RG - \omega^2 LC \tag{2.24}$$

$$\frac{2\alpha}{v_p} = LG + CR \tag{2.25}$$

$$\alpha^2 - \left(\frac{\omega}{v_p}\right)^2 = RG - \omega^2 LC \tag{2.26}$$

Therefore, from equalizing the coefficients of ω^2 and the free term, we get

$$\alpha = \sqrt{RG} \tag{2.27}$$

$$v_p = \frac{1}{\sqrt{LC}} \tag{2.28}$$

Then we substitute in the equation of the imaginary term

$$2\sqrt{RLGC} = LG + CR \tag{2.29}$$

$$2 = \sqrt{\frac{LG}{CR}} + \sqrt{\frac{CR}{LG}} \tag{2.30}$$

The only solution for the above equation is when

$$\frac{LG}{CR} = 1 \tag{2.31}$$

which can be rewritten as

$$\frac{G}{C} = \frac{R}{L} \tag{2.32}$$

The above equation presents the general condition for a dispersionless transmission line. So even for a lossy transmission line, it can be made dispersionless if we ensure the above condition is met.

2.3 Lossless Transmission Line

For a lossless transmission line ($R = 0$ and $G = 0$), $\gamma = j\beta$ and the characteristic impedance is real and is expressed as (from Equation (2.15))

$$Z_0 = \sqrt{\frac{L}{C}} \tag{2.33}$$

For a lossless transmission line of length ℓ that is terminated with a load impedance Z_L at $x = 0$, we can write the following equations to obtain the input impedance:

$$Z_{in}(-\ell) = \frac{V(-\ell)}{I(-\ell)} = \frac{V_f e^{j\beta\ell} + V_r e^{-j\beta\ell}}{V_f e^{-j\beta\ell} - V_r e^{-j\beta\ell}} Z_0 \tag{2.34}$$

$$Z_L = \frac{V(0)}{I(0)} = \frac{V_f + V_r}{V_f - V_r} Z_0 \tag{2.35}$$

Let's define the ratio of reflected to forward wave amplitudes (V_f / V_r) at $x = 0$ as the load reflection coefficient.

$$\rho_L = \frac{V_r}{V_f} \tag{2.36}$$

From Equations (2.35) and (2.36), we can write

$$Z_L = \frac{1 + \rho_L}{1 - \rho_L} Z_0 \tag{2.37}$$

which can also be rewritten as

$$\rho_L = \frac{Z_L - Z_0}{Z_L + Z_0} \tag{2.38}$$

It should be noted that the reflection coefficient $\rho_L = 0$ when $Z_L = Z_0$. In this condition, we say that the transmission line is terminated with a matched load, and hence no reflection occurs at the load end. In other words, the impedance $Z_{in}(0) = Z_0$ and the transmission line has similar impedance as the infinitely long transmission line with no reflections. The deviation of Z_L from Z_0 determines the amount of reflected wave from the load. Now we can write $Z_{in}(-\ell)$ in terms of Z_L and Z_0 as follows:

$$Z_{in}(-\ell) = \frac{1 + \rho_L e^{-j2\beta\ell}}{1 - \rho_L e^{-j2\beta\ell}} Z_0 \tag{2.39}$$

$$Z_{in}(-\ell) = \frac{Z_L \left(1 + e^{-j2\beta\ell}\right) + Z_0 \left(1 - e^{-j2\beta\ell}\right)}{Z_0 \left(1 + e^{-j2\beta\ell}\right) + Z_L \left(1 - e^{-j2\beta\ell}\right)} Z_0 \tag{2.40}$$

which leads to the famous equation

$$Z_{in}(-\ell) = \frac{Z_L + j Z_0 \tan(\beta\ell)}{Z_0 + j Z_L \tan(\beta\ell)} Z_0 \tag{2.41}$$

The above equation is quite interesting and will prove very useful in later chapters. The average power flow in the transmission line in the forward direction is expressed as

$$P_{av} = \frac{1}{2} Re \left(V(x) I^*(x)\right) \tag{2.42}$$

Substituting from (2.11) and (2.12):

$$P_{av}(x) = \frac{1}{2Z_0} Re \left(\left(V_f e^{-\gamma x} + V_r e^{\gamma x}\right)\left(V_f^* e^{-\gamma^* x} - V_r^* e^{\gamma^* x}\right)\right) \tag{2.43}$$

$$= \frac{1}{2Z_0} Re \left(|V_f|^2 e^{-(\gamma+\gamma^*)x} - |V_r|^2 e^{(\gamma+\gamma^*)x} \right.$$
$$\left. + V_f^* V_r e^{(\gamma-\gamma^*)x} - V_f V_r^* e^{-(\gamma-\gamma^*)x}\right) \tag{2.44}$$

$$= \frac{1}{2Z_0} Re \left(|V_f|^2 e^{-2\alpha x} - |V_r|^2 e^{2\alpha x} + V_f^* V_r e^{2j\beta x} - V_f V_r^* e^{-2j\beta x}\right) \tag{2.45}$$

$$= \frac{1}{2Z_0} \left(|V_r|^2 e^{-2\alpha x} - |V_r|^2 e^{2\alpha x}\right) \tag{2.46}$$

$$= \frac{|V_f|^2 e^{-2\alpha x}}{2Z_0} \left(1 - |\rho_L|^2 e^{4\alpha x}\right) \tag{2.47}$$

For a lossless transmission line, this average power flow is reduced to

$$P_{av}(x) = \frac{|V_f|^2}{2Z_0}\left(1 - |\rho_L|^2\right) \tag{2.48}$$

which is independent of x since there is no power loss through the transmission line. From the above equation, we can calculate the input power at the source at $x = -\ell$ and the power delivered to the load at $x = 0$:

$$P_L = P_{av}(0) = \frac{|V_f|^2}{2Z_0}\left(1 - |\rho_L|^2\right) \tag{2.49}$$

$$P_{in} = P_{av}(-\ell) = \frac{|V_f|^2 e^{2\alpha\ell}}{2Z_0}\left(1 - |\rho_L|^2 e^{-4\alpha\ell}\right) \tag{2.50}$$

Therefore, we can express the power transfer efficiency of the transmission line as the ratio of the power delivered to the load to the input power,

$$\eta = \frac{P_L}{P_{in}} = e^{-2\alpha\ell}\frac{1 - |\rho_L|^2}{1 - |\rho_L|^2 e^{-4\alpha\ell}} \tag{2.51}$$

which is dependent on the load reflection coefficient, the transmission line length, and the attenuation constant. Zero efficiency occurs when $|\rho_L| = 1$, where no power is delivered to the load. The maximum power transfer efficiency for a given transmission line length occurs for a matched load with $\rho_L = 0$ and is expressed as

$$\eta_{max} = e^{-2\alpha\ell} \tag{2.52}$$

2.4 Shorted Transmission Line

For the case of shorted transmission line ($Z_L = 0$), Equation (2.41) can be simplified to

$$Z_{in}(-\ell) = jZ_0 \tan(\beta\ell) \tag{2.53}$$

At $\beta\ell = \frac{\pi}{2}$ and its odd multiples, the input impedance of a shorted transmission line is infinity. Since $\beta = \frac{2\pi}{\lambda}$, where λ is the wavelength of the traveling wave, this infinite input impedance condition occurs at $\ell = \frac{\lambda}{4}$ and its odd multiples as illustrated in Figure 2.3. This observation is particularly useful in bias circuits since the transmission line will be acting as a short circuit at DC while it presents an open circuit at the desired frequency. At even multiples of $\ell = \frac{\lambda}{4}$, the input impedance is zero. It is also important to note the input impedance of the shorted transmission line is inductive for $0 < \ell < \frac{\lambda}{4}$, and it becomes capacitive for $\frac{\lambda}{4} < \ell < \frac{\lambda}{2}$. This is particularly useful to simplify the model of the transmission line using a lumped inductor and capacitor

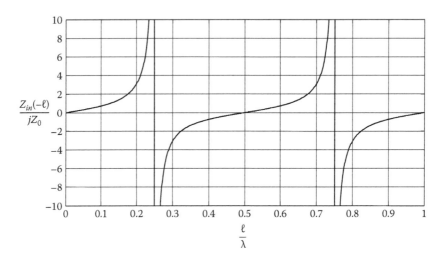

FIGURE 2.3
Input impedance of a lossless transmission line terminated with a short circuit.

for a narrow band of frequencies. An interesting observation about the input impedance shown in Figure 2.3 is the apparent discontinuity at odd multiples of $\ell = \frac{\lambda}{4}$, where the reactance switches from $-\infty$ to ∞, or vice versa. This discontinuity is better understood when we study the lossy transmission line ($\alpha > 0$) and see what happens when α approaches zero. For a lossy shorted transmission line, the input impedance can be written as

$$Z_{in}(-\ell) = \tanh(\gamma \ell) Z_0 = \tanh(\alpha \ell + j\beta \ell) Z_0 = R_{in} + j X_{in} \qquad (2.54)$$

The real and imaginary components of the above expression are plotted in Figure 2.4 for several values of α. It is noted that as ℓ increases from zero toward $\frac{\lambda}{4}$, both input resistance and reactance increase. As ℓ approaches $\frac{\lambda}{4}$, the input resistance continues to increase, but the input reactance peaks at some value of ℓ, and then drops to zero at $\ell = \frac{\lambda}{4}$. At this value of ℓ, the input resistance is at its maximum. As α decreases toward zero, the point of maximum reactance approaches $(\ell, X_{in}) = (\frac{\lambda}{4}, \infty)$, while the maximum input resistance decreases until it becomes zero for a lossless transmission line. It is important to note that for a finite nonzero value of α, the input impedance of the transmission line is always finite, regardless of its length ℓ.

2.5 Voltage Standing Wave Ratio (VSWR)

For a lossless transmission line, the magnitude of the voltage waveform at any point x can be expressed as

$$|V(x)| = |V_f||1 + \rho_L e^{2j\beta x}| = |V_f||1 + |\rho_L| e^{j(\theta + 2\beta x)}| \qquad (2.55)$$

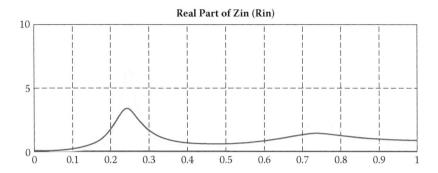

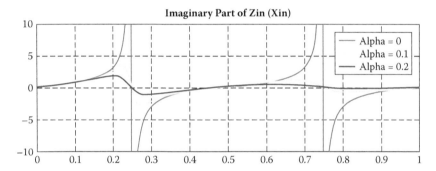

FIGURE 2.4
Input resistance and reactance of a lossy transmission line terminated with a short circuit.

where θ is the phase of the load reflection coefficient. From the above expression, the maximum and minimum voltage magnitudes along the transmission line can be calculated as

$$V_{max} = |V_f|(1 + |\rho_L|) \tag{2.56}$$
$$V_{min} = |V_f|(1 - |\rho_L|) \tag{2.57}$$

and the ratio of maximum to minimum voltage magnitudes is called the voltage standing wave ratio (VSWR) and is expressed as

$$VSWR = \frac{1 + |\rho_L|}{1 - |\rho_L|} \tag{2.58}$$

The importance of the VSWR comes from the ease of measurement since it is a voltage ratio. The load impedance can be indirectly calculated by measuring the maximum and minimum voltage magnitudes along the transmission line and the location of the voltage minimum. The VSWR can be used to calculate $|\rho_L|$, while the location of the voltage minimum (ℓ_{min}) is used to calculate θ from the following equations:

$$|\rho_L| = \frac{VSWR - 1}{VSWR + 1} \tag{2.59}$$

$$\theta = \pi + 2\beta\ell_{min} \tag{2.60}$$

2.6 Smith Chart

The Smith chart is a graphical tool to calculate the impedance as a function of the reflection coefficient. Equation (2.6) can be generalized for the impedance and reflection coefficient at any point x on the transmission line:

$$Z(x) = \frac{1 + \rho(x)}{1 - \rho(x)} Z_0 \tag{2.61}$$

To simplify our equations, we normalize the impedance as $Z_n(x) = Z(x)/Z_0$.

$$Z_n(x) = \frac{1 + \rho(x)}{1 - \rho(x)} \tag{2.62}$$

Let's define $Z_n(x) = R_n + j X_n$ and $\rho(x) = a + jb$; then

$$R_n + j X_n = \frac{1 + a + jb}{1 - a - jb} = \frac{(1 + a + jb)(1 - a + jb)}{(1 - a)^2 + b^2} \tag{2.63}$$

Therefore, we can write R_n and X_n in terms of a and b:

$$R_n = \frac{1 - a^2 - b^2}{(1 - a)^2 + b^2} \tag{2.64}$$

$$X_n = \frac{2b}{(1 - a)^2 + b^2} \tag{2.65}$$

We can rewrite Equation (2.64) as follows:

$$\left(a - \frac{R_n}{1 + R_n} \right)^2 + b^2 = \frac{1}{(1 + R_n)^2} \tag{2.66}$$

which is an equation of a circle in the ρ-plane, centered at $(\frac{R_n}{1 + R_n}, 0)$ with a radius of $\frac{1}{1 + R_n}$. We can also rewrite Equation (2.65) as follows:

$$(a - 1)^2 + \left(b - \frac{1}{X_n} \right)^2 = \frac{1}{X_n^2} \tag{2.67}$$

which is an equation of another circle in the ρ-plane, with a center at $(1, \frac{1}{X_n})$ and a radius of $\frac{1}{X_n}$. The transformation from horizontal lines in the z-plane into circles in the ρ-plane is illustrated in Figure 2.6. The graphical representations of Equations (2.66) and (2.67) are illustrated in Figures 2.5 and 2.6. In Figure 2.5, the vertical R_n lines in the z-plane are transformed into circles in the ρ-plane with centers along the a-axis. All R_n circles are tangent to the $a = 1$ line at the point $(a, b) = (1, 0)$, where $|Z_n| = \infty$ according to Equation (2.62). It is important to recognize the special circles in the ρ-plane. The circle $R_n = 0$ is

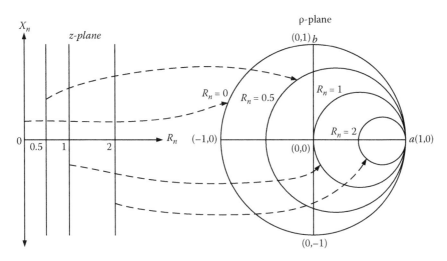

FIGURE 2.5
Transformation from vertical lines in the z-plane into circles in the ρ-plane.

equivalent to $|\rho| = 1$, which is called the unit circle. The unit circle respresents the boundary between negative values of R_n outside the unit circle ($|\rho| > 1$ where reflected power from the load is larger than incident power) and positive values of R_n inside the unit circle ($|\rho| < 1$ where reflected power is less than incident power). For $R_n = 0$, the circle in the ρ-plane passes through the origin, which represents a matched termination ($|\rho| = 0$). As $R_n \to \infty$,

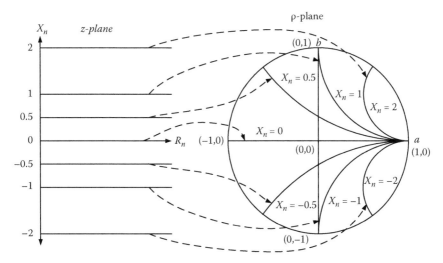

FIGURE 2.6
Transformation from horizontal lines in the z-plane into circles in the ρ-plane.

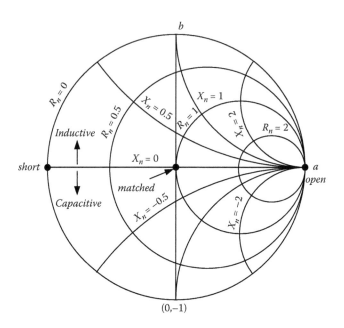

FIGURE 2.7
Basic Smith chart showing X_n and R_n lines.

the R_n circles shrink toward the point $(a, b) = (1, 0)$. For passive elements, $0 < R_n < \infty$, and we always operate inside the unit circle ($|\rho| < 1$). Therefore, Smith charts only show the ρ-plane inside the unit circle.

On the other hand, the horizontal X_n lines in the z-plane are transformed into circles in the ρ-plane with centers along the vertical line $a = 1$. For passive elements, only the part of X_n circles inside the unit circle of the ρ-plane is shown. All X_n circles are tangent to the a-axis at the point $(a, b) = (1, 0)$. For $X_n = 0$, the circle has an infinite radius according to Equation (2.67) and reduces to a point on the a-axis. For inductive impedances where $X_n > 0$, the circles reside in the top half of the ρ-plane. On the other hand, for capacitive impedances where $X_n < 0$, circles reside in the bottom half of the ρ-plane.

The above observations from Figures 2.5 and 2.6 are summarized in Figure 2.7, which is considered the basic Smith chart. Both R_n and X_n circles are shown on the same chart so that we can read Z_n and ρ values instantaneously. It is important to note that the short-circuit ($Z_n = 0$) and open-circuit ($|Z_n| = \infty$) terminations are located at leftmost and rightmost points on the unity circle periphery. The matched point ($Z_n = 1$ and $\rho = 0$) is located at the center of the unit circle.

As we move toward the generator, the reflection coefficient $\rho(x)$ can be calculated as follows:

$$\rho(x) = \frac{V_r e^{\gamma x}}{V_f e^{-\gamma x}} = \rho_L e^{2\gamma x} \tag{2.68}$$

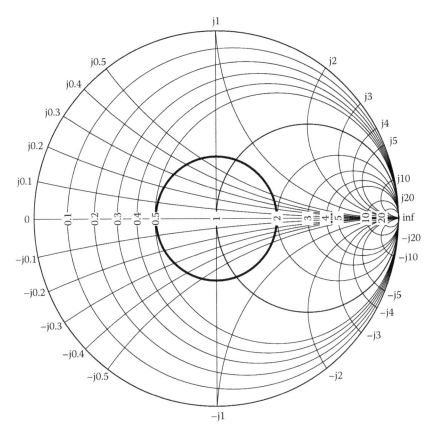

FIGURE 2.8
ρ vs. normalized length for a lossless transmission line.

For a lossless transmission line, $\rho(x) = \rho_L e^{2j\beta x}$. The magnitude of ρ is independent of x and phase is $4\pi \frac{x}{\lambda} = 4\pi x_n$, where $x_n = \frac{x}{\lambda}$ is the normalized distance from the load. Hence, moving toward the generator (in the $-x$ direction) on the transmission line is equivalent to moving counter-clockwise on a circle with its center at the origin of the Smith chart. We spin $360°$ around the origin of the Smith chart for every $\frac{\lambda}{2}$ of the transmission line. This is illustrated in Figure 2.8 for the case of a normalized load impedance $Z_n = 0.5$. The reflection coefficient at the load $\rho_L = \rho(0) = \frac{1}{3}$.

For the case of a lossy transmission line, $\rho(x) = \rho_L e^{2\alpha x} e^{2j\beta x}$, and the magnitude of α drops exponentially as we move toward the generator (away from the load). This is illustrated in Figure 2.9 for the case of $Z_n = 0.5$. For an infinitely long transmssion line, $\rho(x)$ drops to zero as x approaches $-\infty$. However, this does not mean that higher power is delivered to the load for longer transmission line. It simply means that higher power is transmitted from the generator into the transmission line and higher losses in the transmission line itself. This is evident from Equation (2.51).

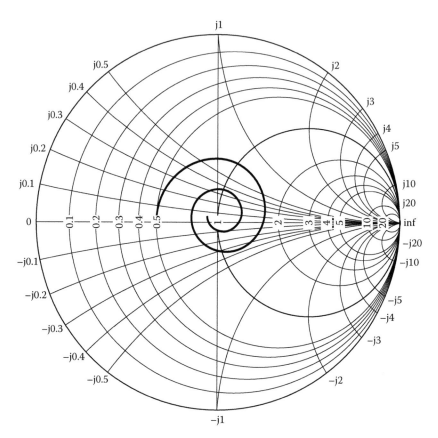

FIGURE 2.9
ρ vs. normalized length for a lossy transmission line.

The Smith chart that we have discussed so far is also called the impedance Smith chart. Alternatively, one can derive the admittance Smith chart by writing

$$Y_n(x) = Z_n(x) = \frac{1 - \rho(x)}{1 + \rho(x)} \tag{2.69}$$

Figure 2.10 illustrates the admittance Smith chart where constant-conductance and constant-susceptance circles are drawn. The constant-conductance circles are all tangent to the $a = -1$ line on the ρ-plane at the point $(a, b) = (-1, 0)$. The constant-susceptance circles are all tangent to the a-axis at the same point. Note that the locations of the short circuit ($Y_n = \infty$) and open circuit ($Y_n = 0$) in the ρ-plane are the same as they are in the impedance Smith chart. Note also that negative susceptance cirles are at the top half of the admittance chart, while positive susceptance circles are at the bottom half. Therefore, the top half still represents inductive elements, while the bottom half represents capacitive elements, as in the impedance chart.

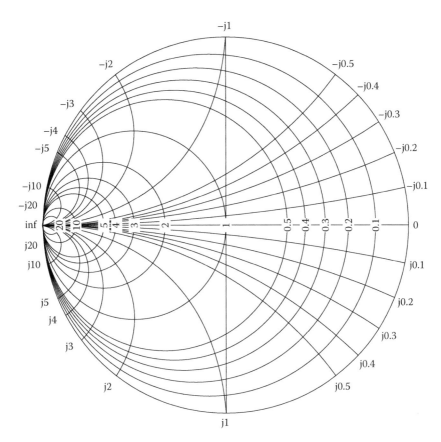

FIGURE 2.10
Admittance Smith chart.

Bibliography

1. R. E. Collin. (1966). *Foundations for Microwave Engineering*. New York: McGraw-Hill, 1966.
2. U. L. Rohde, G. D. Vendelin, and A. M. Pavio. (1990). *Microwave Circuit Design Using Linear and Nonlinear Techniques*. New York: John Wiley, 1990.
3. D. M. Pozar. (1997). *Microwave Egineering. 2nd Edition*. New York: John Wiley, 1997.

3

Trigger Mode Distributed Wave Oscillator

Many of today's electronic systems use an oscillator circuit as a high frequency signal source. In the last decade or two, many transmission line-based new traveling or standing wave oscillator techniques have been introduced as a very high frequency signal source with the availability of oscillation phases at the tap points along the line. Availability of such GHz-range, high-resolution oscillation phases is one of the most significant advantages of these oscillators compared to their LC-based lumped counterparts. LC-based lumped ones use two reactive components, an inductor and a capacitor, to create a resonant circuit, in an ideal case indefinitely transferring the energy from one to the other. However, in reality, the loss mechanisms associated with these reactive devices (can be modeled as resistance (R) or transconductance (G) elements) require active amplifying circuitry to compensate for these losses. The well-known classical implementation for such an active compensation circuit is a negative resistance circuit formed by cross-coupled active devices. The metal oxide semiconductor field effect transistor (MOSFET) implementation of this configuration is shown in Figure 3.1. The resultant oscillation frequency depends on the inductance and the capacitance values and can be written as

$$f_{osc} = \frac{1}{2\pi \sqrt{L_{tank}C_{tank}}} \tag{3.1}$$

Similarly, distributed counterparts can be constructed using transmission lines. A transmission line is, in general, parallel running conductors separated by a dielectric material. Microstrip line (Figure 3.2), coplanar waveguide (Figure 3.3), coplanar strip line (Figure 3.4), and differential coplanar waveguide (Figure 3.5) are some of the most common transmission line structures. Although any of these structures can be used to construct an oscillator, the differentially symmetric ones are more favorable since the opposite phases of a signal are already available (coplanar strip line and differential coplanar waveguide).

Since these transmission lines effectively represent a distributed LC structure, an oscillator similar to a lumped LC tank oscillator can be formed as shown in Figure 3.6. In this figure, L_0, C_0, R_0, and G_0 represent inductance per unit length, capacitance per unit length, resistance per unit length, and conductance per unit length for a differential transmission line stretching in the z-direction. The inductance per unit length and capacitance per unit length determine the phase velocity of the wave propagating. The phase velocity of

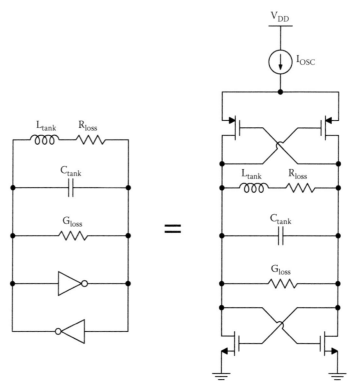

FIGURE 3.1
Lumped LC tank oscillator.

a wave is as follows:

$$v = \frac{1}{\sqrt{L_0 C_0}} \tag{3.2}$$

where L_0 and C_0 are inductance per unit length and capacitance per unit length, respectively. Then, for a given total length of transmission line, the

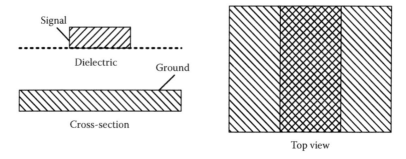

FIGURE 3.2
Microstrip line.

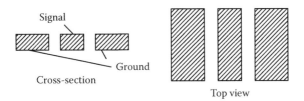

FIGURE 3.3
Coplanar waveguide.

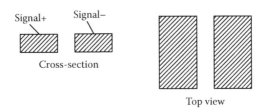

FIGURE 3.4
Coplanar strip line.

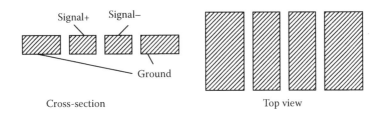

FIGURE 3.5
Differential coplanar waveguide.

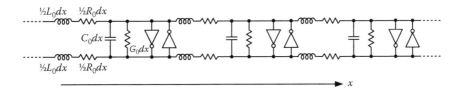

FIGURE 3.6
Distributed oscillator structure using transmission lines.

oscillation frequency can be calculated as

$$f_{osc} = \frac{1}{\sqrt{L_{tot}C_{tot}}} \tag{3.3}$$

where L_{tot} and C_{tot} are the total inductance and total capacitance along the transmission line. Again, cross-coupled active amplifiers are used to compensate for the conductor and substrate losses. Thanks to the distributed nature of these transmission line oscillators, multiple phases of an oscillation are available along the transmission line, whereas only two 180° opposite phases are available in the case of lumped LC tank oscillators. Distributed wave oscillators, rotary traveling wave oscillators, and standing wave oscillators are different classes of existing transmission line-based oscillators all utilizing the distributed LC nature of a transmission line structure. These existing topologies will be touched upon briefly in Sections 3.1 to 3.3. Section 3.4 introduces a new topology, trigger mode distributed wave oscillator (TMDWO) [7], and discusses its advantages and disadvantages compared to existing topologies. Section 3.6 presents a test structure and the related measurement results of the proposed technique, and Section 3.7 is the conclusion.

3.1 Commonly Used Wave Oscillator Topologies

One of the earliest traveling wave oscillator inventions was distributed wave oscillator [2–4,9,10,16]. Figure 3.7 shows a simplified distributed oscillator structure. The actual shape can be in any closing geometric form bringing point A to the vicinity of point B so that a dashed AC coupled connection can

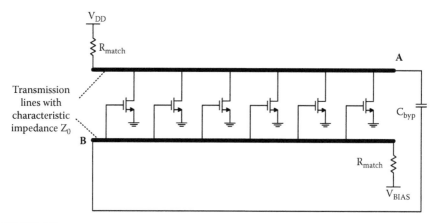

FIGURE 3.7
Basic distributed oscillator structure.

be obtained using a capacitor C_{byp}. The reflections resulting from the mismatch of the biasing resistor, R_{match}, to the line impedance, Z_0, can be a significant source of disturbance in the steady-state oscillation waveforms. This effect, together with an additional nonideality due to the bypass capacitor C_{byp}, is the main drawback of the technique.

3.2 Rotary Traveling Wave Oscillator (RTWO)

Another transmission line oscillator technique, rotary traveling wave oscillator technique, avoids this disadvantage by direct cross-coupling of the end points with an additional cost of odd symmetry introduced by this crossing of the transmission lines (Figure 3.8) [8,15]. The single-wire closed-loop structure of an RTWO limits the disturbances to one crossover, which can still be significant, especially at high frequencies. Once enough gain is provided, there is no latch-up danger for the technique, since it utilizes a single-line DC-coupled closed-loop structure. A practical circular layout of the RTWO is shown in Figure 3.9, where the end points are brought together to implement the crossover connection. Mutlimode oscillation may occur in RTWO with frequencies that satisfy the periodic boundary condition in the continuous limit [1]. The mode frequencies may not be exact multiples of the fundamental frequency due to dispersion (see Section 2.2). Even modes, where the two lines are excited with the same signal, are suppressed with the cross-coupled inverters along the RTWO. Therefore, only odd modes may exist, where the two lines are excited with opposite phases. Odd modes are supported by the cross-coupled inverters along the RTWO. So, the differential coupled line can be simplified with a single-ended model. To analyze the odd modes of oscillation in the RTWO, let's first consider the poles of the short-circuited lossless transmission line of length ℓ. The driving point admittance of such a

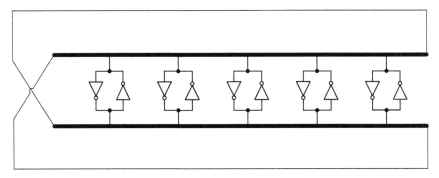

FIGURE 3.8
Rotary traveling wave oscillator.

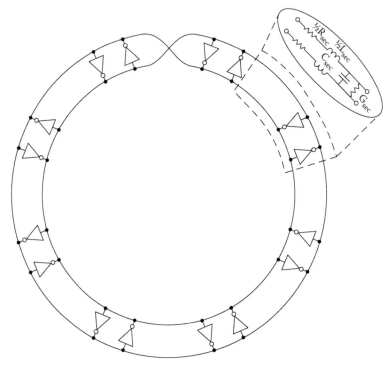

FIGURE 3.9
Rotary traveling wave oscillator: circular layout.

transmission line can be obtained from (2.53) as

$$Y_{in} = \frac{1}{Z_{in}(-\ell)} = \frac{-j}{Z_0} \cot(\beta\ell) \tag{3.4}$$

The poles of the short-circuit driving point admittance occur at frequencies where $Y_{in} = \infty$ or $\beta\ell = n\pi$, where n is the pole number. Using Equation (2.18), we obtain the pole locations as follows:

$$p_n = \pm j\frac{n\pi}{\ell\sqrt{LC}} = \pm j\frac{n\pi}{\sqrt{L_{sec}C_{sec}}} \tag{3.5}$$

where $L_{sec} = \ell L$ and $C_{sec} = \ell C$ are the total inductance and capacitance of the transmission line section, respectively. Equation (3.5) gives an infinite number of possible oscillation modes. Realistically, however, the maximum oscillation frequency is determined by the transmission line losses or the periodic loading of the transmission line.

The RTWO is loaded periodically with the cross-coupled inverters as shown in Figure 3.9, where C_{sec} represents the lumped capacitance of the transmission line, the parasitic capacitances of the cross-coupled inverters, and any other capacitaces needed for biasing or tuning the RTWO. If the RTWO is approximated as a chain of transmission line sections as shown in Figure 3.9, the poles in the case of lossless network (assuming $R_{sec} = 0$ and $G_{sec} = 0$) are [1]:

$$p_n = \pm \frac{2j}{\sqrt{L_{sec}C_{sec}}} \sin\left(\frac{n\pi}{2N}\right) \tag{3.6}$$

where N is the number of transmission line sections in the RTWO. For large values of N, Equation (3.6) reduces to Equation (3.5), due to the fact that RTWO is more distributed in this case. The highest-order mode occurs when $n = N$ when the pole frequency is

$$p_n = \pm \frac{2j}{\sqrt{L_{sec}C_{sec}}} = \omega_C \tag{3.7}$$

This mode frequency represents the cutoff frequency of the resonator, which limits the maximum possible rise time, and hence limits the minimum oscillator phase noise that can be obtained.

It is neccessary to calculate the quality factor (Q) of the resonator in order to calculate the phase noise and determine the size of the cross-coupled inverters to guarantee oscillation. By definition, the quality factor is determined by

$$
\begin{aligned}
Q &= 2\pi \frac{energy\ stored}{Energy\ dissipated\ in\ one\ cycle} \\
&= 2\pi f_0 \frac{energy\ stored}{average\ power\ dissipation}
\end{aligned} \tag{3.8}
$$

where the numerator is the energy stored in the L_0 and C_0 distributed components, while the denominator is the average power dissipated in the R_0 and G_0 components. Therefore, we can obtain an expression of the transmission quality factor as

$$Q = \omega_0 \frac{\frac{1}{2}L_0 I_p^2 + \frac{1}{2}C_0 V_p^2}{\frac{1}{2}R_0 I_p^2 + \frac{1}{2}G_0 V_p^2} \tag{3.9}$$

where I_p and V_p are the peaks of current and voltage waveforms, respectively. Substituting $V_p = Z_0 I_p$, we get

$$Q = \omega \frac{L_0 + C_0 Z_0^2}{R_0 + G Z_0^2} = \omega \frac{L_0/Z_0 + C_0 Z_0}{R_0/Z_0 + G Z_0} \tag{3.10}$$

For a low loss transmission line, we can substitute $Z_0 = \sqrt{\frac{L_0}{C_0}}$; therefore, we obtain

$$Q = \frac{2\omega\sqrt{L_0 C_0}}{R_0/Z_0 + G Z_0} = \frac{\omega}{\omega_{3db}} \tag{3.11}$$

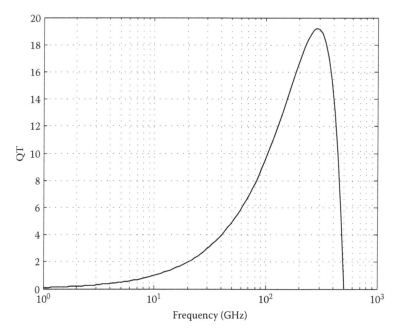

FIGURE 3.10
Q_T of periodically loaded transmission line with $\omega_{3db} = 10$ GHz and $\omega_c = 500$ GHz.

where

$$\omega_{3db} = \frac{R_0/Z_0 + G Z_0}{2\sqrt{L_0 C_0}} \tag{3.12}$$

The quality factor equation (3.11) is valid only for distributed transmission line. For periodically loaded transmission line, Equation (3.11), the quality factor can be modified as

$$Q_T = Q \left(1 - \left(\frac{\omega}{\omega_c} \right)^2 \right) \tag{3.13}$$

$$= \frac{\omega}{\omega_{3db}} \left(1 - \left(\frac{\omega}{\omega_c} \right)^2 \right) \tag{3.14}$$

Therefore, the above equation is very important in determining the quality factor at different modes of oscillation, and hence the possibility of oscillation. Equation (3.14) is plotted in Figure 3.10 for $\omega_{3db} = 10$ GHz and $\omega_c = 500$ GHz. The maximum Q_T can be obtained by differentiating Equation (3.14) with respect to ω and equalizing it to zero:

$$\frac{d Q_T}{d\omega} = 0 = \frac{1}{\omega_{3db}} - \frac{3}{\omega_{3db}} \left(\frac{\omega}{\omega_c} \right)^2 \tag{3.15}$$

Therefore, the frequency of the maximum Q_T is

$$\omega_{Q_{Tmax}} = \frac{\omega_c}{\sqrt{3}} \tag{3.16}$$

which is substituted in Equation (3.14) to obtain the maximum Q_T:

$$Q_{Tmax} = \frac{2}{3\sqrt{3}} \frac{\omega_c}{\omega_{3db}} \tag{3.17}$$

3.3 Standing Wave Oscillator (SWO)

Standing wave oscillators are another group of transmission line oscillators that would utilize transmission line structures [5,6,12,13]. Standing waves are formed by superimposing the forward and backward traveling waves on the same transmission medium simultaneously. The two basic standing wave oscillator topologies, $\lambda/4$ SWO and $\lambda/2$ SWO, are shown in Figures 3.11 and 3.12, respectively [12]. A $\lambda/2$ SWO is basically a combination of two $\lambda/4$ SWOs around a center symmetry point, with the fundamental operating principle staying the same. In this type of oscillator, the differential transmission line structure is driven by a cross-coupled amplifier pair at one end, whereas the other end is shorted. The waves created at the amplifier end are reflected back at the short end, causing a reverse propagating wave along the transmission line. In the steady state, the forward and reverse waves coexist, creating a standing wave along the line. This would imply amplitude variations in the oscillation phases along the line, gradually diminishing and eventually reaching zero at the short end. To derive an expression for the standing wave along the transmission line in Figure 3.11, we recall Equation (2.11):

$$V(x) = V_f e^{-\gamma x} + V_r e^{\gamma x} \tag{3.18}$$

Then we substitute $\gamma = j\beta$ for a lossless transmission line and use Equation (2.36):

$$V(x) = V_f \left(e^{-j\beta x} + \rho_L e^{j\beta x} \right) \tag{3.19}$$

For a shorted transmission line, $\rho_L = -1$ according to Equation (2.38); therefore:

$$V(x) = V_f \left(e^{-j\beta x} - e^{j\beta x} \right) = -2jV_f sin(\beta x) \tag{3.20}$$

The above equation indicates a standing wave that has zero amplitude at the shorted load ($x = 0$) and maximum amplitude at the cross-coupled amplifier pair ($x = \frac{\lambda}{4}$). The phase of the voltage waveform is independent of x. In other words, the sum of forward traveling wave (where phase increases with x) and reverse traveling wave (where phase decreases with x) results in a standing

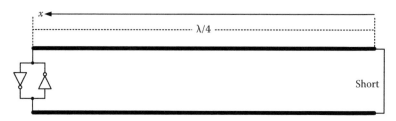

FIGURE 3.11
λ/4 standing wave oscillator.

wave where phase is constant along the transmission line. We can write the expression of the current along the transmission line from Equation (2.12):

$$I(x) = \frac{V_f}{Z_0}\left(e^{-j\beta x} + e^{j\beta x}\right) = \frac{2V_f}{Z_0}\cos(\beta x) \tag{3.21}$$

which indicates maximum current amplitude at the shorted load and zero ampliude at the cross-coupled amplifier pair.

Circular standing wave oscillator (CSWO) is another standing wave technique that would not require any reflection mechanism, but rather a circular symmetry to create reverse propagating waves along the transmission line medium [5,6]. As shown in Figure 3.13, the energy is injected into a closed-loop transmission line structure equally and travels symmetrically along the ring in clockwise and counterclockwise directions. These counter-traveling waves create standing waves with an amplitude profile as shown in Figure 3.14. The energy is injected at two opposite points (*A* and *B*) with additional dashed connections to force the main mode. Additionally, at least one of the quiet ports has to be shorted to prevent any latch-up problems. This also reduces this structure to a single-line structure.

All of these standing wave oscillator structures have a critical drawback of amplitude variations, which would limit their usage to a very limited set of applications. The oscillation phases corresponding to the quiet ports would not even exist, compromising the main advantage of transmission line oscillators.

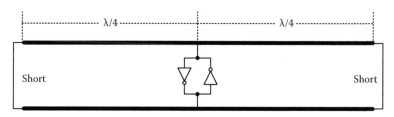

FIGURE 3.12
λ/2 standing wave oscillator.

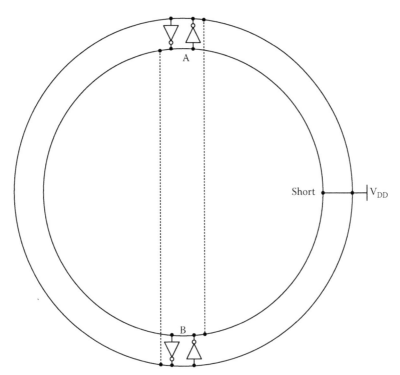

FIGURE 3.13
CSWO structure.

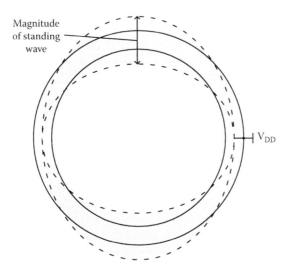

FIGURE 3.14
Amplitude profile along the CSWO structure.

3.4 TMDWO Topology

The trigger mode distributed wave oscillator described in this section proposes an alternative approach for a wave oscillator that can provide unique advantages compared to existing prior techniques described in the preceding section. Figure 3.15 shows the diagram of the proposed traveling wave technique. In this technique, an oscillation is triggered in two independent transmission lines, each carrying opposite phases of the oscillation at any particular location along the parallel running differential transmission lines. The waves traveling on these two independent lines would possess the same propagation characteristic since the lines and the loading corresponding to each line are identical and symmetric. This means that once successfully created, the opposite phases of an oscillation signal can propagate indefinitely together with the help of cross-coupled amplifiers that are finely distributed along the lines compensating for the losses. These cross-coupled amplifier unit cells should be distributed in maximally symmetric fashion, resulting in a smooth traveling wave without any amplitude or phase distortion. These inverting amplifiers use the signal in one of the lines as booster for the opposing phase traveling in the other line. However, the difficulty arises with the initial existence of these opposite oscillation phases in the independent identical lines. Since the constituent transmission lines are two independent conductors, there is no mechanism to guarantee traveling wave buildup. Thus, during the power-up, the system would latchup even before any oscillation buildup (i.e., one of the lines would be pulled up to V_{DD} and the other one to GND due to the cross-coupled amplifiers). This condition is overcome by using another auxiliary oscillator that injects multiple opposite phases into the corresponding phase locations along the loop during the power-up. This would also ensure an oscillation only in the fundamental mode.

The closed shape of these parallel-running differentially triggered transmission lines, which is shown to be circular, can take any symmetrical geometric form ending up with symmetrical injection points corresponding to the available triggering phases. Figure 3.16 shows the direction of these resulting opposite traveling waves, while Figure 3.17 shows the time domain waveforms for some of the oscillation phases corresponding to the various locations along the transmission line. Another unique feature of the technique that relaxes the routing requirements of the available phases in the actual physical layout is that it can provide all of the oscillation phases in both of the two independent conductors, the same phase being available at the two opposite sides with respect to the axis of symmetry. Hence, the structure can be viewed as triggering two identical independent transmission line loops in opposite phases so that each one of them uses the other as a sustainer.

Figure 3.15 illustrates a four-phase trigger mode traveling wave oscillator. Any well-known quadrature oscillator type, such as a four-stage differential ring voltage-controlled oscillator (VCO) or a quadrature LC tank VCO, can be

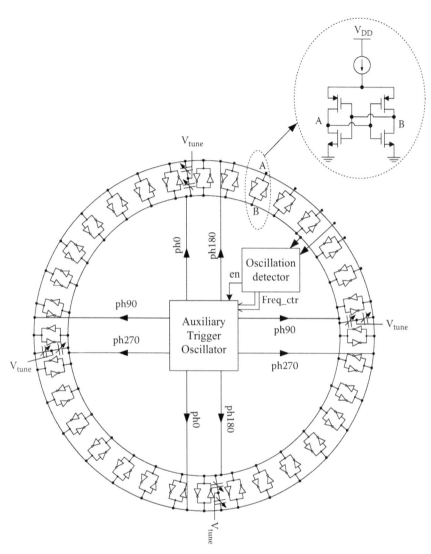

FIGURE 3.15
TMDWO circuit diagram.

used as a triggering auxiliary oscillator. The schematic of the four-stage differential ring VCO, including an individual delay element, is shown in Figure 3.18.

Every delay cell in this VCO is connected to a voltage-controlled current source to provide frequency tuning. An oscillator with 8, 16, or more number of phases can also be used for this purpose, routing the available phases to their corresponding locations along the transmission lines. The quadrature

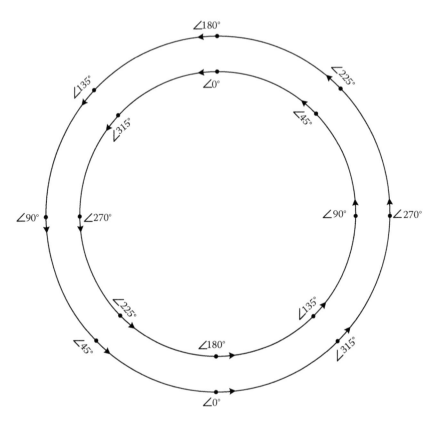

FIGURE 3.16
Opposite wave propagation in TMDWO conductors.

phases for this particular case are denoted as *ph0, ph90, ph180*, and *ph270*. An important requirement for this triggering oscillator is that its tuning range should cover the estimated frequency range of the traveling wave oscillator. This triggering oscillator is powered up first, injecting the quadrature signals into the corresponding quadrature locations on the transmission line ring. As this close-by oscillation is injected into the ring, the supply for the cross-coupled amplifiers, VDD_ring, starts to ramp up. As the ring supply ramps up, the weak injection oscillation is amplified by these cross-coupled amplifiers, tracking the supply and effectively preventing the lines to latch up. Finally, after a successful oscillation buildup, the detector circuit detects the oscillation and powers down the triggering auxiliary oscillator to save power, and thereafter the resultant traveling wave would sustain itself unless a long-lasting power glitch causes it to latch up. In the case of such an occasion resulting in latch-up, the detector circuit reinitiates the start-up sequence to rebuild the oscillation. However, in order for the defined start-up sequence to be successful, the triggering oscillation frequency should be in the close

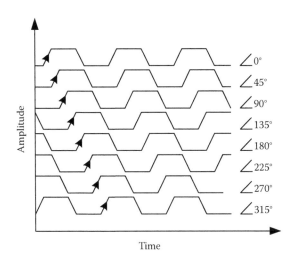

FIGURE 3.17
Time domain representation of some of the oscillation phases on the lines.

vicinity of the actual traveling wave frequency that would end up in the lines. Since the parameters governing the oscillation frequencies of both oscillators are process dependent, they can vary considerably. In order to guarantee a successful triggering in varying environments and process conditions, the mentioned power-up sequence is repeated for a wide range of frequencies as the triggering oscillator frequency is swept across a wide frequency range. In every sweep step, the oscillation detector looks for an oscillation. In case no oscillation is detected, the detector circuit updates the triggering frequency to the next step by a frequency control word, freq-ctr, and restarts the supply ramp sequence. Once a successful oscillation is detected, the sweep process

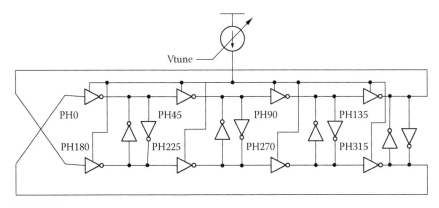

FIGURE 3.18
A four-stage trimmable differential VCO as a trigger oscillator.

is terminated and the triggering oscillator is powered down. Although the technique might sound inferior considering the additional triggering circuitry required, the even symmetry proposed by the technique can provide a very high phase accuracy that is very difficult to obtain at very high frequencies with the classical wave oscillator techniques. This is mainly due to the fact that there is no source of asymmetry, such as odd number of line crossings or real termination impedances that may not match the line impedance accurately. Moreover, since each of these independent transmission lines corresponds to a full lap of traveling wave rather than half, using the same total conductor length, one can obtain double-oscillation frequency compared to existing distributed wave oscillators. This implies that much higher frequencies can be obtained using this technique. Regarding the triggering circuitry that may be considered to be a disadvantage of this circuit, the overhead is insignificant. The area cost of an extra triggering circuitry in an integrated circuit implementation is minimal, since the area of a ring oscillator is an order of magnitude smaller than the area of the main oscillator. Power is also not a concern since the triggering circuitry is powered down after triggering the sequence, consuming no power during normal operation. It should also be noted that since the triggering is applied in a symmetric distributed fashion, and the amplifiers are also finely distributed in a very symmetric fashion, there is source of reflection or any distinctive preferred way for the energy to split. Thus, the wave is forced into traveling mode rather than standing mode. Standing wave oscillation generally is not a preferred mode of operation due to severe amplitude variations in the oscillation phases. The distributed varactors shown in Figure 3.15 constitute a fraction of the total capacitance on the line providing a tuning range for the TMTWO, and hence enabling its use in phase-locked loops. The control voltage, Vtune, changes the total amount of capacitor on the line, thus determining the frequency of the traveling wave.

3.5 Phase Noise in Traveling Wave Oscillators

The main advantage of traveling wave oscillators over the traditional CMOS ring oscillators is that phases are generated with transmission line delays rather than transistor delays. Therefore, we expect to have better phase noise performance in traveling wave-based oscillators. In this section, we aim to derive the phase noise of the SWO and RTWO [11]. We will start with the phase noise in SWO, from which we can derive the phase noise in RTWO.

3.5.1 Phase Noise in SWO

The $\lambda/2$ SWO discussed in Section 3.3 can be decomposed into two parallel $\lambda/4$ SWOs, as shown in Figure 3.11. To simplify the analysis, we start with $\lambda/4$

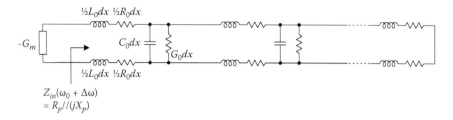

FIGURE 3.19
Model of the λ/4 SWO.

SWO phase noise calculation, and then we can derive the phase noise of the
$\lambda/2$ SWO. The $\lambda/4$ SWO is modeled as shown in Figure 3.19.

The input impedance of the $\lambda/4$ SWO, at an offset frequency $\Delta\omega$ from the
oscillation frequency ω_0, can be calculated from (2.53) by setting $\ell = \lambda/4$.
Therefore:

$$Z_{in}(\omega_0 + \Delta\omega) = j Z_0 \tan(\beta\lambda/4) \tag{3.22}$$

where β is the propagation constant and is a function of the input signal
frequency. Therefore, in this equation, we cannot write $\beta = 2\pi/\lambda$ because
we are interested in finding the value at $\omega_0 + \Delta\omega$ rather than ω_0. We recall
Equation (2.18), which is written as

$$\beta(\omega_0 + \Delta\omega) = (\omega_0 + \Delta\omega)\sqrt{LC} \tag{3.23}$$

Hence, we obtain the expression of Z_{in} as a function of $\Delta\omega$ and ω_0:

$$Z_{in}(\omega_0 + \Delta\omega) = j Z_0 \tan\left(\left(1 + \frac{\Delta\omega}{\omega_0}\right)\omega_0\sqrt{LC}\lambda/4\right) \tag{3.24}$$

$$= j Z_0 \tan\left(\left(1 + \frac{\Delta\omega}{\omega_0}\right)\pi/2\right) \approx -j Z_0\frac{2\omega_0}{\pi\,\Delta\omega} \tag{3.25}$$

The frequency of oscillation is related to the propagation velocity and the
wavelength as

$$\omega_0 = \frac{2\pi v_p}{\lambda} = \frac{\pi v_p}{2\ell_q} \tag{3.26}$$

where ℓ_q is the length of the quarter-wave shorted transmission line. Substituting from Equation (2.28), we obtain

$$\omega_0 = \frac{\pi}{2\ell_q\sqrt{LC}} = \frac{\pi}{2\sqrt{L_{sec}C_{sec}}} \tag{3.27}$$

where L_{sec} and C_{sec} are the total inductance and capacitance of the quarter-
wave transmission line, including the input capacitances of the cross-coupled

inverters. Substituting from Equation (3.27) into Equation (3.25), we get the magnitude of the input impedance as

$$|Z_{in}(\omega_0 + \Delta\omega)| \approx \frac{4\omega_0^2 L_{sec}}{\pi^2 \Delta\omega} \tag{3.28}$$

To obtain the quality factor of the resonator, the resonator loss is represented with a shunt resistance R_p at the transmission line. To obtain the value of R_p for a distributed shorted transmission line, we recall Equation (2.54) for a lossy transmission line:

$$Z_{in}(\omega_0 + \Delta\omega) = \tanh(\gamma\ell)\, Z_0 = \tanh(\alpha\ell + j\beta\ell)\, Z_0$$

$$= \frac{1 - e^{-2\alpha\ell - 2j\beta\ell}}{1 + e^{-2\alpha\ell - 2j\beta\ell}} Z_0$$

$$= \frac{1 - e^{-2\alpha\ell - j(1 + \frac{\Delta\omega}{\omega_0})\pi}}{1 + e^{-2\alpha\ell - j(1 + \frac{\Delta\omega}{\omega_0})\pi}} Z_0$$

$$= \frac{1 + e^{-2\alpha\ell - j\frac{\Delta\omega}{\omega_0}\pi}}{1 - e^{-2\alpha\ell - j\frac{\Delta\omega}{\omega_0}\pi}} Z_0$$

$$\approx \frac{2Z_0}{+2\alpha\ell + j\frac{\Delta\omega}{\omega_0}\pi} = \frac{Z_0}{\alpha\frac{\lambda}{4} + j\frac{\Delta\omega}{\omega_0}\frac{\pi}{2}} = \frac{1}{\frac{1}{R_p} + \frac{j}{X_p}} \tag{3.29}$$

Therefore, we can obtain the values of R_p and X_p as follows:

$$R_p = Z_0 / \left(\alpha\frac{\lambda}{4}\right) \tag{3.30}$$

$$X_p = Z_0 \frac{2\omega_0}{\pi\,\Delta\omega} \tag{3.31}$$

Note that the expression for X_p agrees with Equation (3.25). By substituting the value of α from Equation (2.17), we obtain the general expression for R_p:

$$R_p = \frac{8Z_0}{(\frac{R}{Z_0} + G Z_0)\lambda} \tag{3.32}$$

Now we can find the $3dB$ bandwidth of the input impedance Z_{in} by writing

$$\left| \frac{Z_{in}(\omega_0 + \omega_{3dB})}{Z_{in}(\omega_0)} \right| = \frac{1}{\sqrt{2}} \tag{3.33}$$

It is easier to write it in terms of input admittance:

$$\left| \frac{Y_{in}(\omega_0)}{Y_{in}(\omega_0 + \omega_{3dB})} \right| = \frac{1}{\sqrt{2}} \tag{3.34}$$

$$\left| \frac{1/R_p}{1/R_p + j/X_p(\omega_0 + \omega_{3dB})} \right| = \frac{1}{\sqrt{2}} \tag{3.35}$$

Therefore, at the 3dB frequency we have

$$X_p(\omega_0 + \omega_{3dB}) = R_p = Z_0 \frac{2\omega_0}{\pi \Delta \omega} \tag{3.36}$$

where Equation (3.31) is used. Now we can write the expression for the 3dB bandwidth as

$$\omega_{3dB} = Z_0 \frac{2\omega_0}{\pi R_p} \tag{3.37}$$

Finally, we obtain the quality factor of the resonator as

$$Q = \frac{\omega_0}{2\omega_{3dB}} = \frac{\omega_0}{2Z_0 \frac{2\omega_0}{\pi R_p}} = \frac{\pi R_p}{4Z_0} \tag{3.38}$$

By substituting R_p from Equation (3.30), we get

$$Q = \frac{\pi}{\alpha \lambda} \tag{3.39}$$

To study the phase noise of $\lambda/4$ SWO, assume the NMOS cross-coupled pair shown in Figure 3.21 is used. The cross-coupled pair is biased through the short-circuit side of the $\lambda/4$ transmission line. Assuming current-limited operation, the differential current flowing into the transmission line is ideally a square wave of amplitude I_B. The fundamental component of this current waveform is $\frac{4}{\pi} I_B$. Therefore, the amplitude of oscillation at the fundamental frequency is expressed as

$$V_p = \frac{4}{\pi} I_B R_p \tag{3.40}$$

In this oscillator structure, there are three sources of noise: the transmission line loss represented by R_p, the tail transistor noise, and the commutating pair noise. Next, we will study the phase noise due to these individual noise sources.

3.5.1.1 Noise of Transmission Line

Transmission line distributed loss is lumped into one element, R_p, which is expressed in terms of the transmission line parameters in Equation (3.32). The power spectrum density of the thermal current noise of the equivalent resistor R_p is

$$\overline{i_{nRp}^2} = 4kT/R_p \tag{3.41}$$

where k is Boltzmann's constant, and T is the absolute temperature. As suming half of this noise contributes to phase noise (and the other half to amplitude noise), we can write the phase noise due to resonator noise as

$$
\begin{aligned}
\mathcal{L}_{resonator} &= \frac{|Z_{in}(\omega_0 + \Delta\omega)|^2 \frac{\overline{i_{ri}^2}}{2}}{\frac{V_p^2}{2}} \\
&= \frac{4kT \, |Z_{in}(\omega_0 + \Delta\omega)|^2}{R_p V_p^2}
\end{aligned}
\tag{3.42}
$$

At offset frequencies larger than ω_{3dB}, we can approximate the input impedance to the shorted transmission line from Equation (3.29) as

$$
|Z_{in}(\omega_0 + \Delta\omega)| = \left| \frac{Z_0}{\alpha\frac{\lambda}{4} + j\frac{\Delta\omega}{\omega_0}\frac{\pi}{2}} \right| \approx \frac{2\omega_0}{\pi\Delta\omega} Z_0
\tag{3.43}
$$

Therefore, the phase noise due to resonator resistive losses expressed from Equation (3.42) in terms of the offset frequency is

$$
\mathcal{L}_{resonator} = \frac{4kT \, Z_0^2}{R_p V_p^2} \left(\frac{2\omega_0}{\pi\Delta\omega} \right)^2
\tag{3.44}
$$

Note that the phase noise due to oscillator resistive losses drops as 20 dB/decade of the offset frequency $\Delta\omega$.

3.5.1.2 Tail Transistor Noise

The noise of the tail transistor is commutated by the cross-coupled pair as shown in Figure 3.22. The cross-coupled pair acts as the commutating pair in active up-conversion mixers. It flips the polarity of the tail transistor noise at the differential output every half-oscillation period. The differential output current noise is expressed as

$$
\begin{aligned}
i_{no-tail} &= i_{np} - i_{nm} \\
&= i_n S(t)
\end{aligned}
\tag{3.45}
$$

where $S(t)$ is a square wave that models the commutating action of the cross-coupled pair. Assuming ideal hard switching, $S(t)$ is approximated as a square wave with 50% duty cycle and unity magnitude. For wideband white noise of the tail transistor, the commutated output noise is also white and has the same power spectral density at the tail transistor noise. Therefore, the power spectral density of the differential output noise current is expressed as

$$
\overline{i_{no-tail}^2} = 4kT\gamma g_m B
\tag{3.46}
$$

where γ is the channel noise factor, which is equal to 2/3 for long channel devices. For deep submicron technologies, γ ranges from 2 to 3 or more.

Similar to the resonator thermal noise, we can write the phase noise due to the tail therm noise as

$$\mathcal{L}_{tail} = \frac{4kT\gamma g_{mB} Z_0^2}{V_p^2} \left(\frac{2\omega_0}{\pi \Delta\omega} \right)^2 \tag{3.47}$$

We can express the transconductance of the tail transistor as

$$g_{mB} = \frac{2I_B}{V_{eff}} \tag{3.48}$$

From Equations (3.40) and (3.48), we obtain

$$g_{mB} = \frac{\pi V_p}{2R_p V_{eff}} \tag{3.49}$$

Therefore, we can rewrite Equation (3.47) as

$$\mathcal{L}_{tail} = \frac{4kT Z_0^2}{R_p V_p^2} \frac{\pi \gamma V_p}{2V_{eff}} \left(\frac{2\omega_0}{\pi \Delta\omega} \right)^2 \tag{3.50}$$

3.5.1.3 Cross-Coupled Pair Noise

The noise of the cross-coupled transistors only exists when both transistors are ON, and this occurs around the zero crossing of the differential resonator voltage at the cross-coupled pair location. Therefore, the output noise due to the cross-coupled pair is modulated with resonator voltage, and this type of noise is referred to as cyclostationary [14]. Figure 3.20(a) and (b) shows the input voltages and the differential output current of the cross-coupled pair. The peak transconductance of the cross-coupled pair is approximated with the expression

$$G_{mp} = \frac{2I_B}{\Delta V} \tag{3.51}$$

where ΔV is the differential voltage required to steer the current completely into one device of the cross-coupled pair. An approximation of the transconductance waveform of the cross-coupled pair vs. time is shown in Figure 3.20(c). The resultant noise waveform is shown in Figure 3.20(d), which indicates the cyclostationary property of the output noise. The average power spectral density of the cyclostationary noise at the output is expressed as

$$\overline{i_{no-cc-pair}^2} = 4kT\gamma G_m \frac{2\Delta T}{T_0} \tag{3.52}$$

where ΔT is the period when both cross-coupled pair transistors are ON, which can be expressed in terms of ΔV as

$$\Delta T = \frac{\Delta V}{S} \tag{3.53}$$

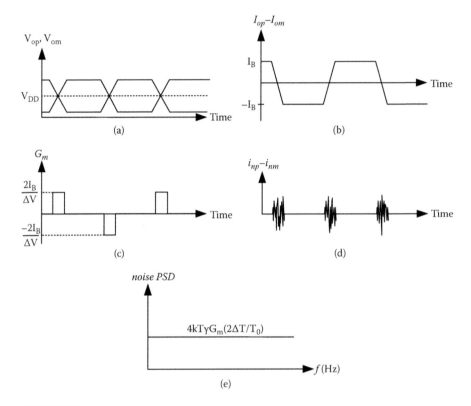

FIGURE 3.20
(a) Resonator voltage waveforms. (b) Differential output current. (c) Cross-coupled pair transconductance. (d) Differential output noise due to cross-coupled pair. (e) Noise power spectral density.

where S is the slope of the differential voltage waveform at the zero crossing. Therefore, we can write the power spectral density as

$$\overline{i^2_{no-cc-pair}} = 4kT\gamma G_m \frac{2\Delta V}{S} \frac{\omega_0}{2\pi} = 4kT\gamma \frac{2I_B\omega_0}{\pi S} \tag{3.54}$$

Since the cross-coupled pair noise is generated around the zero-crossing point of the differential voltage, the whole noise power contributes to phase noise (unlike noise originating from the tail transistor or the resonator losses). Therefore, we can express the phase noise due to the cross-coupled pair thermal noise as

$$\mathcal{L}_{cc-pair} = \frac{4kT Z_0^2}{V_p^2} \frac{4\gamma I_B\omega_0}{\pi S} \left(\frac{2\omega_0}{\pi \Delta\omega}\right)^2 \tag{3.55}$$

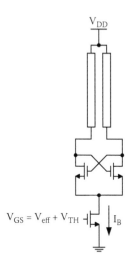

FIGURE 3.21
Commutation of the tail transistor thermal noise by the cross-coupled pair.

By combining Equations (3.44), (3.50), and (3.55), we obtain the overall thermal phase noise for the $\lambda/4$ SWO as

$$\mathcal{L}_{\text{SWO}-\lambda/4}(\Delta\omega) = \frac{4kT Z_0^2}{R_p V_p^2}\left(\frac{2\omega_0}{\pi \Delta\omega}\right)^2\left[1 + \frac{\pi \gamma V_p}{2V_{\text{eff}}} + \frac{4\gamma I_B R_p \omega_0}{\pi S}\right] \qquad (3.56)$$

Now we are in a position to calculate the phase noise of a $\lambda/2$ SWO by considering the cross-coupled pair of transconductors loaded with two parallel

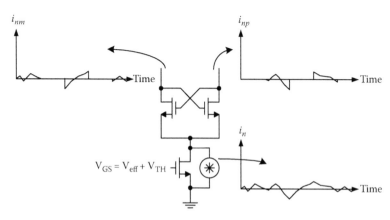

FIGURE 3.22
Commutation of the tail transistor thermal noise by the cross-coupled pair.

$\lambda/4$ transmission lines. This is obtained by replacing R_p with $R_p/2$, V_p with $V_p/2$, and S with $S/2$ in Equation (3.56):

$$\mathcal{L}_{SWO-\lambda/2}(\Delta\omega) = \frac{64kT\,Z_0^2}{R_p\,V_p^2}\left(\frac{2\omega_0}{\pi\,\Delta\omega}\right)^2\left[1 + \frac{\pi\gamma\,V_p}{4V_{eff}} + \frac{4\gamma\,I_B\,R_p\omega_0}{\pi\,S}\right] \quad (3.57)$$

3.5.2 Phase Noise in RTWO

Similar to the SWO, in order to calculate the phase noise of the RTWO, we need to calculate the impedance seen by the cross-coupled pair at the injection point. To carry out this calculation, let's consider the four-section RTWO (Figure 3.23). The wave propagation is assumed to be in the counterclockwise direction. Therefore, the voltage and current phase relationship at different injection points can be described with the following equations:

$$V_n = \begin{cases} V_{n-1}e^{-j\beta\lambda_0/2N} & \text{for} \quad n = 1, 2, \ldots, 2N-1 \\ V_{2N}e^{-j\beta\lambda_0/2N} & \text{for} \quad n = 0 \end{cases} \quad (3.58)$$

$$I_n = \begin{cases} I_{n-1}e^{-j\beta\lambda_0/2N} & \text{for} \quad n = 1, 2, \ldots, 2N-1 \\ I_{2N}e^{-j\beta\lambda_0/2N} & \text{for} \quad n = 0 \end{cases} \quad (3.59)$$

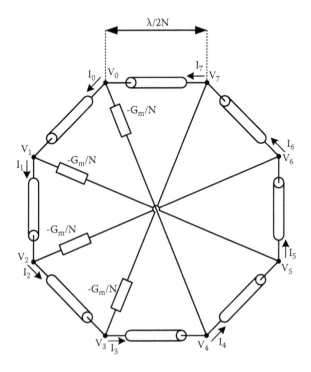

FIGURE 3.23
Four-section RTWO ($N = 4$) with counterclockwise wave propagation.

where N is the number of cross-coupled transcondutances ($N = 4$ in Figure 3.23), and λ_0 represents the wavelength of the propagating wave at the frequency of oscillation. Recall from Equation (2.18) that the propagation constant can be related to the operating frequency as

$$\beta = \omega \sqrt{LC} \tag{3.60}$$

where L and C are the inductance and capacitance per unit length of the transmission line, respectively. At the resonance frequency, the overall phase shift around the RTWO is equal to 2π, and the phase shift in each section is $2\pi/2N$. Therefore, we can write

$$\beta_0 \lambda_0 / 2N = 2\pi / 2N \tag{3.61}$$

Therefore, the propagation constant is related to the wavelength at the frequency of oscillation as

$$\beta_0 = \frac{2\pi}{\lambda_0} \tag{3.62}$$

We can write the Kirchhoff current law (KCL) equation at each of the injection nodes as

$$I_n = I_{n-1} e^{-j\beta\lambda_0/2N} e^{-\alpha\lambda_0/2N} + \frac{G_m}{N}(V_n - V_{n-N}) \tag{3.63}$$

Combining Equations (3.59) and (3.63), we obtain

$$I_n = I_n e^{-\alpha\lambda_0/2N} + \frac{G_m}{N}(V_n - V_{n-N}) \tag{3.64}$$

At the oscillation frequency, $V_n = -V_{n-N}$, so Equation (3.64) can be simplified to

$$I_n = I_n e^{-\alpha\lambda_0/2N} + \frac{2G_m}{N}V_n(x) \tag{3.65}$$

Therefore, the required minimum total transconductance G_m to ensure traveling wave oscillation is

$$G_m = \frac{NI_n}{2V_n}\left(1 - e^{-\alpha\lambda_0/2N}\right) \approx \frac{\alpha\lambda_0}{4Z_0} \tag{3.66}$$

Therefore, the required total transconductance of an N-section RTWO is independent of N.

Now if we assume a current I_{in} with frequency $\omega_0 + \Delta\omega$ is injected at node n as shown in Figure 3.24, we can substitute $\beta = \beta_0\left(1 + \frac{\Delta\omega}{\omega_0}\right)$ into Equation (3.63) and obtain

$$I_{in} + I_n = I_{n-1} e^{-j\beta_0\lambda_0\left(1 + \frac{\Delta\omega}{\omega_0}\right)/2N} e^{-\alpha\lambda_0/2N} + \frac{G_m}{N}(V_n - V_{n-N}) \tag{3.67}$$

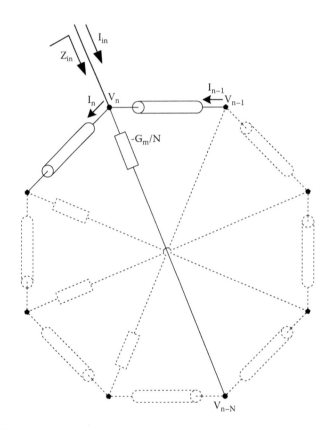

FIGURE 3.24
Calculating the input impedance of the RTWO at an injection point.

At an offset frequency $\Delta\omega$, we can express I_{n-1} and V_{n-N} as

$$V_{n-N} = V_n e^{-j\beta_0 \left(1+\frac{\Delta\omega}{\omega_0}\right)\lambda_0/2} = V_n e^{-j\pi\left(1+\frac{\Delta\omega}{\omega_0}\right)} = -V_n e^{-j\pi\frac{\Delta\omega}{\omega_0}} \qquad (3.68)$$

$$I_{n-1} = I_n e^{-j(2N-1)\beta_0\lambda_0\left(1+\frac{\Delta\omega}{\omega_0}\right)/2N} \qquad (3.69)$$

Substituting in Equation (3.67), we obtain the simplified equation

$$I_{in} + I_n = I_n e^{-j2\pi\frac{\Delta\omega}{\omega_0}} e^{-\alpha\lambda_0/2N} + \frac{G_m}{N} V_n \left(1 + e^{-j\pi\frac{\Delta\omega}{\omega_0}}\right) \qquad (3.70)$$

By dividing both sides by I_n and using the fact that $V_n/I_n = Z_0$ and $I_{in} = V_n/Z_{in}$, we obtain

$$\frac{Z_0}{Z_{in}} + 1 = e^{-j2\pi\frac{\Delta\omega}{\omega_0}} e^{-\alpha\lambda_0/2N} + \frac{G_m}{N} Z_0 \left(1 + e^{-j\pi\frac{\Delta\omega}{\omega_0}}\right) \qquad (3.71)$$

For small $\frac{\Delta\omega}{\omega_0}$, the above equation can be simplified to

$$\frac{Z_0}{Z_{in}} = -j2\pi\frac{\Delta\omega}{\omega_0} - \alpha\lambda_0/2N + \frac{G_m}{N}Z_0\left(2 - j\pi\frac{\Delta\omega}{\omega_0}\right) \tag{3.72}$$

By combining Equations (3.66) and (3.72), we obtain

$$\frac{Z_0}{Z_{in}} = -j2\pi\frac{\Delta\omega}{\omega_0}\left(1 + \frac{\alpha\lambda_0}{8N}\right) \tag{3.73}$$

Finally, the closed-form expression for the magnitude of Z_{in} is

$$|Z_{in}| = Z_0\frac{\omega_0}{2\pi\,\Delta\omega\left(1 + \frac{\alpha\lambda_0}{8N}\right)} \tag{3.74}$$

Since all noise sources are modeled differentially between V_n and V_{n-N} as shown in Figure 3.25, we need to derive the differential impedance between as

$$|Z_{ind}| = \left|\frac{V_n - V_{n-N}}{I_{ind}}\right| = \left|\frac{Z_0}{-j2\pi\frac{\Delta\omega}{\omega_0}\left(1 + \frac{\alpha\lambda_0}{8N}\right)}\left(1 + e^{-j\pi\frac{\Delta\omega}{\omega_0}}\right)\right| \approx \frac{Z_0}{\pi\left(1 + \frac{\alpha\lambda_0}{8N}\right)}\frac{\omega_0}{\Delta\omega} \tag{3.75}$$

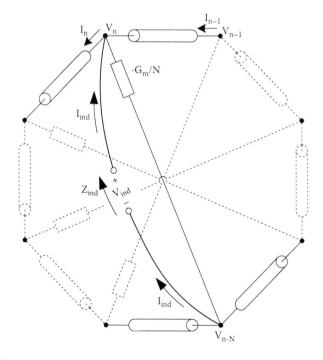

FIGURE 3.25
Calculating the differential input impedance of an N-section RTWO between nodes n and $n - N$.

Note that the input impedance of the RTWO is at least two times smaller than the SWO in Equation (3.43), which corresponds to better phase noise. To complete the phase noise calculation, we need to find an expression for the peak voltage in terms of the resonator loss. From the condition of oscillation, the effective differential resistance seen by the cross-coupled transconductance between nodes n and $n - N$ is expressed as

$$R_p = \frac{N}{G_m} \tag{3.76}$$

Using Equation (3.66), we obtain

$$R_p = \frac{4NZ_0}{\alpha \lambda_0} \tag{3.77}$$

The amplitude of oscillation can then be calculated from Equations (3.40) and (3.77) as

$$V_p = \frac{4}{\pi} \frac{I_B}{N} R_p = \frac{16 I_B Z_0}{\pi \alpha \lambda_0} \tag{3.78}$$

where I_B is the total current in the N cross-coupled transconductances. Similar to the phase noise calculation in SWO, the phase noise of an N-section RTWO due to the noise originating from one section (including noise from the transmission line of this section, tail thermal noise, and cross-coupled transistors noise) is expressed as

$$\mathcal{L}_{\text{RTWO-section}}(\Delta\omega) = \frac{4kT Z_0^2}{R_p V_p^2} \left(\frac{1}{1 + \frac{\alpha\lambda_0}{8N}} \frac{\omega_0}{\pi \Delta\omega} \right)^2 \left[1 + \frac{\pi \gamma V_p}{2V_{\text{eff}}} + \frac{4\gamma I_B R_p \omega_0}{N\pi S} \right] \tag{3.79}$$

The overall phase noise from the contributions of all N sections is expressed as

$$\mathcal{L}_{\text{RTWO}}(\Delta\omega) = \frac{4NkT Z_0^2}{R_p V_p^2} \left(\frac{1}{1 + \frac{\alpha\lambda_0}{8N}} \frac{\omega_0}{\pi \Delta\omega} \right)^2 \left[1 + \frac{\pi \gamma V_p}{2V_{\text{eff}}} + \frac{4\gamma I_B R_p \omega_0}{N\pi S} \right] \tag{3.80}$$

By eliminating I_B from Equation (3.80) using Equation (3.78), we obtain the overall phase noise of N-section RTWO as

$$\mathcal{L}_{\text{RTWO}}(\Delta\omega) = \frac{4NkT Z_0^2}{R_p V_p^2} \left(\frac{1}{1 + \frac{\alpha\lambda_0}{8N}} \frac{\omega_0}{\pi \Delta\omega} \right)^2 \left[1 + \frac{\pi \gamma V_p}{2V_{\text{eff}}} + \frac{\gamma V_p \omega_0}{S} \right] \tag{3.81}$$

It is interesting to check the effect of the number of sections N on the overall phase noise for the same total length λ_0 and total current I_B. To find this dependence, we need to eliminate any parameter in Equation (3.81) that is dependent on N. The only parameter that depends on N is R_p, which is proportional to N according to Equation (3.77). Therefore, we get

$$\mathcal{L}_{RTWO}(\Delta\omega) = \frac{\alpha\lambda_0 kTZ_0}{V_p^2}\left(\frac{1}{1+\frac{\alpha\lambda_0}{8N}}\frac{\omega_0}{\pi\Delta\omega}\right)^2\left[1+\frac{\pi\gamma V_p}{2V_{eff}}+\frac{\gamma V_p\omega_0}{S}\right] \qquad (3.82)$$

Thus, phase noise of the RTWO is weakly dependent on the number of section N. In fact, even for $N = 1$, the term $\alpha\lambda_0/8$ is expected to be much smaller than unity ($\alpha\lambda_0$ is the total attenuation along the RTWO length, which is expected to be less than unity). Therefore, the phase noise of the RTWO is approximately given by the following simplified expression:

$$\mathcal{L}_{RTWO}(\Delta\omega) = \frac{\alpha\lambda_0 kTZ_0}{V_p^2}\left(\frac{\omega_0}{\pi\Delta\omega}\right)^2\left[1+\frac{\pi\gamma V_p}{2V_{eff}}+\frac{\gamma V_p\omega_0}{S}\right] \qquad (3.83)$$

3.5.3 Phase Noise in Differential Wave Oscillator (DWO)

Differential wave oscillators are a class of traveling wave oscillators where two differential waves are sustained between two loops as illustrated in Figure 3.26. There are three differential modes of oscillation in this type of oscillator. These modes are clockwise traveling wave, counterclockwise traveling wave, and nontraveling wave. One mode can be forced to exist by design. For example, in the TMDWO in Figure 3.15, the triggering circuit forces the wave

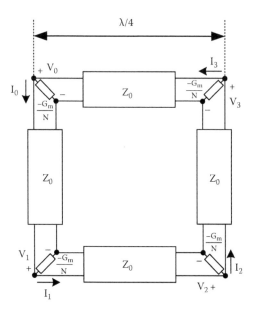

FIGURE 3.26
Four-section DWO ($N = 4$) with counterclockwise wave propagation.

to travel initially in the counterclockwise direction. Assuming counterclockwise wave propagation in Figure 3.26, we can write the differential voltage and current waveforms as follows:

$$V_n = \begin{cases} V_{n-1}e^{-j\beta\lambda_0/N} & for \quad n = 1, 2, \ldots, N-1 \\ V_N e^{-j\beta\lambda_0/N} & for \quad n = 0 \end{cases} \tag{3.84}$$

$$I_n = \begin{cases} I_{n-1}e^{-j\beta\lambda_0/N} & for \quad n = 1, 2, \ldots, N-1 \\ I_N e^{-j\beta\lambda_0/N} & for \quad n = 0 \end{cases} \tag{3.85}$$

where $N = 4$ for the case in Figure 3.26. We can write the KCL equation at node n as

$$I_n = I_{n-1}e^{-j\beta\lambda_0/N}e^{-\alpha\lambda_0/N} + \frac{G_m}{N}V_n \tag{3.86}$$

and by combining Equations (3.85) and (3.86), we obtain

$$I_n = I_n e^{-\alpha\lambda_0/N} + \frac{G_m}{N}V_n \tag{3.87}$$

Therefore, we obtain the minimum total transconductance to ensure oscillation as

$$G_m = \frac{N}{Z_0}\left(1 - e^{-\alpha\lambda_0/N}\right) \approx \frac{\alpha\lambda_0}{Z_0} \tag{3.88}$$

Similar to the analysis of the RTWO, we can calculate the differential input impedance at the injection point n as illustrated in Figure 3.27 at an offset frequency $\Delta\omega$ from the oscillation frequency. We can write the KCL equation as follows:

$$I_{in} + I_n = I_{n-1}e^{-j\beta_0\lambda_0\left(1+\frac{\Delta\omega}{\omega_0}\right)/2N}e^{-\alpha\lambda_0/2N} + \frac{G_m}{N}V_n \tag{3.89}$$

The current I_{n-1} is phase delayed from I_n by $N-1$ sections; therefore, we can write

$$I_{n-1} = I_n e^{-j(N-1)\beta_0\lambda_0\left(1+\frac{\Delta\omega}{\omega_0}\right)/N} \tag{3.90}$$

Substituting from Equation (3.90) into Equation (3.89):

$$I_{in} + I_n = I_n e^{-j2\pi\frac{\Delta\omega}{\omega_0}}e^{-\alpha\lambda_0/N} + \frac{G_m}{N}V_n \tag{3.91}$$

By dividing both sides by I_n and using the fact that $V_n/I_n = Z_0$ and $I_{in} = V_n/Z_{in} = I_n Z_0/Z_{in}$, we obtain

$$\frac{Z_0}{Z_{in}} + 1 = e^{-j2\pi\frac{\Delta\omega}{\omega_0}}e^{-\alpha\lambda_0/N} + \frac{G_m}{N}Z_0 \approx 1 - j2\pi\frac{\Delta\omega}{\omega_0} - \alpha\lambda_0/N + \frac{G_m}{N}Z_0 \tag{3.92}$$

Therefore, by combining Equations (3.88) and (3.92), we obtain the differential input impedance as

$$Z_{in} = j\frac{Z_0}{2\pi}\frac{\omega_0}{\Delta\omega} \tag{3.93}$$

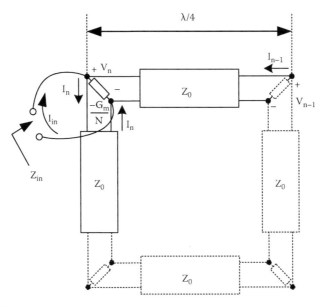

FIGURE 3.27
Calculation of the differential input at the differential injection point n.

The differential resistance seen by the cross-coupled transconductor due to resonator loss is equal to the inverse of the steady-state value of transconductance:

$$R_p = \frac{N}{G_m} = \frac{N Z_0}{\alpha \lambda_0} \tag{3.94}$$

Hence, the peak oscillation voltage is written as

$$V_p = \frac{4}{\pi} \frac{I_B}{N} R_p = \frac{4 I_B Z_0}{\pi \alpha \lambda_0} \tag{3.95}$$

Therefore, we can write the overall phase noise of the DWO as

$$
\begin{aligned}
\mathcal{L}_{\mathrm{DWO}}(\Delta\omega) &= \frac{4NkT Z_0^2}{R_p V_p^2} \left(\frac{\omega_0}{2\pi \Delta\omega} \right)^2 \left[1 + \frac{\pi \gamma V_p}{2 V_{\mathit{eff}}} + \frac{4\gamma I_B R_p \omega_0}{N\pi S} \right] \\
&= \frac{4\alpha \lambda_0 kT Z_0}{V_p^2} \left(\frac{\omega_0}{2\pi \Delta\omega} \right)^2 \left[1 + \frac{\pi \gamma V_p}{2 V_{\mathit{eff}}} + \frac{4\gamma I_B R_p \omega_0}{N\pi S} \right]
\end{aligned} \tag{3.96}
$$

3.6 Experimental Results

The technique can be utilized to generate traveling waves with frequencies ranging from tens of MHz in printed circuit board-level discrete implementations to more than 100 GHz in an integrated circuit environment. A board-level

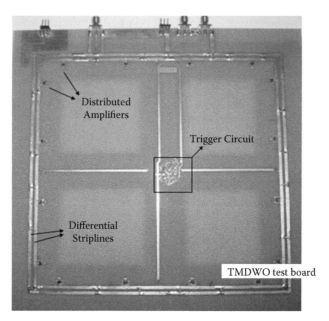

FIGURE 3.28
A board-level discrete implementation for the proposed TMDWO technique.

test structure was designed to prove the concept (Figure 3.28). High-speed CMOS dual-inverter ICs were distributed along the transmission lines as cross-coupled amplifiers. The measured triggering waveforms as well as the resultant oscillation waveforms of this 73 MHz board-level implementation are shown in Figure 3.29.

As predicted by the technique, no oscillation is observed and the lines were latched up unless an injection mechanism was applied by the auxiliary trigger oscillator. Figure 3.30 shows the spectrum of the TMDWO in unlocked open-loop configuration. In order to observe a successful oscillation buildup, the

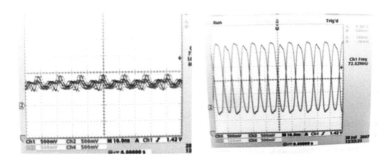

FIGURE 3.29
Measured waveforms of a TMTWO: triggering waveform (left) and resultant oscillation waveform (right).

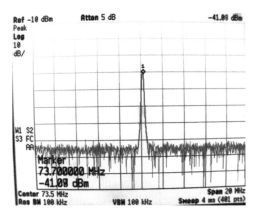

FIGURE 3.30
Open-loop output spectrum of the 73 MHz TMDWO.

triggering signal frequency had to be in the ±%5 vicinity of the final traveling wave frequency.

3.7 Conclusion

Some of the recent innovations in transmission line-based wave oscillators have been discussed. A new transmission line oscillator structure, TMDWO, is introduced, elaborating on its advantages and disadvantages compared to the existing techniques. The concept was verified with a relatively low frequency (73 MHz) board-level setup to guide the GHz-range on-chip implementations. Its main disadvantage, an additional auxiliary oscillator requirement, is estimated to be insignificant in an IC implementation since the target for this additional circuitry is only the triggering of the main oscillator. Hence, the die area to be allocated can be relatively small. Moreover, it can be powered down immediately after start-up.

Bibliography

1. Y. Chen and K. D. Pedrotti. Rotary traveling-wave oscillators, analysis and simulation. *IEEE Transactions on Circuits and Systems I*, 58(1):77–87, Jan. 2011.
2. L. Divina and Z. Skvor. The distributed oscillator at 4 GHz. *Microwave Theory and Techniques, IEEE Transactions on Microwave Theory and Techniques* 46(12):2240–2243, Dec. 1998.
3. N. Goto and E. Honma. U.S. Patent 5640112, Clock signal distributing system, June 1997.

4. H. Wu and S.A. Hajimiri. U.S. Patent 6529085B2, Tunable distributed voltage-controlled oscillator, March 2005.

5. D. Ham and W. Andress. A circular standing wave oscillator. In *Solid-State Circuits Conference, 2004. Digest of Technical Papers. ISSCC. 2004 IEEE International Solid-State Circuits Conference, ISSCC 2004*, (1):380–533, Feb. 2004.

6. D. Ham, W.F. Andress, and Y. Liu, Methods and apparatus based on coplanar striplines, US Patent 7091802, Aug. 2006.

7. D. Ismailov. Trigger-mode distributed wave oscillator system, US Patent 7,741,921, June 2010.

8. J. Wood, Low noise oscillator, US Patent 7,218,180, March 2007.

9. I.P. Kaminow and J. Liu. Propagation characteristics of partially loaded two-conductor transmission line for broadband light modulators. *Proceedings of the IEEE*, 51(1):132–136, Jan. 1963.

10. B. Kleveland, C.H. Diaz, D. Vock, L. Madden, T.H. Lee, and S.S. Wong. Monolithic cmos distributed amplifier and oscillator. In *Solid-State Circuits Conference, 1999. Digest of Technical Papers. ISSCC. 1999 IEEE International*, Solid-State Circuits Conference, ISSCC 1999, 70–71, Feb. 1999.

11. S. Osman Koji Takinami, R. Walsworth, and S. Beccue. Phase-noise analysis in rotary traveling-wave oscillators using simple physical model. *IEEE Transactions on Microwave Theory and Techniques*, 58(6):1465–1474, June 2010.

12. L. Leyten and A.G. Wagemans, Balanced oscillator having a short circuited quarter-wave paired line, US patent 6,342,820, Jan. 2002.

13. F. O' Mahony, C.P. Yue, M. Horowitz, and S.S. Wong. 10ghz clock distribution using coupled standing-wave oscillators. In Solid-State Circuits Conference, ISSCC 2003, *IEEE International*, (1):428, Feb. 2003.

14. J. Phillips and K. Kundert. Noise in mixers, oscillators, samplers, and logic. In *IEEE Custom Integrated Circuits Conference*, 431–438, 2000.

15. J. Wood, T.C. Edwards, and S. Lipa. Rotary traveling-wave oscillator arrays: a new clock technology. *Solid-State Circuits, IEEE Journal of*, Solid-State Circuits Conference, ISSCC 2001, 36(11):1654–1665, Nov. 2001.

16. H. Wu and A. Hajimiri. Silicon-based distributed voltage-controlled oscillators. *Solid-State Circuits, IEEE Journal of*, 36(3):493–502, March 2001.

4

Force Mode Distributed Wave Oscillator

In this chapter, an electronic oscillator circuit that can provide accurate multiple phases of an oscillation and a phased-array power amplification system are introduced. The chapter presents the first measurement results of the ideas presented in [1]. An oscillation can be formed across two independent conductor loops by introducing an additional force mechanism across each of the individual lines, forming a differential transmission medium for the oscillation wave. Two different approaches are proposed in Section 4.1 regarding the application of this additional force mode that prevent the two complementary conductors latching up. In the delay-based first method, the signals at certain points along each line are amplified through inverting amplifiers and applied to the corresponding opposite phase points along the same line. The force application point is calculated depending on the connecting line delays. In the symmetry-based second method, the differential structure is shaped symmetrically to create a meeting point for the opposite phase points of each conductor in the center, thus eliminating the inaccuracy of the connecting line delays. Section 4.2 goes over single-ended version of the oscillator as a radiating element, which is called force-mode distributed wave antenna. Also in Section 4.2, a unidirectional single-conductor version of force-mode distributed wave oscillator (FMDWO) is combined with a secondary pickup coil to form a distributed oscillator amplifier. The proposed oscillator-transformer scheme converts the unidirectional oscillator current in the primary coil into a large-voltage swing in the second coil driving the antenna. By modulating the oscillation frequency with baseband data using varactors or other capacitive tuning elements, a topology that serves as a radio frequency (RF) transmitter is obtained. By segmentation of the total loop into $2N$ sections, distributed multiple local transformer loops are used to utilize the same circulating oscillator current with $360/2N$ phase difference. The system consisting of multiple sections each driving a separate antenna thus serves as a beam-forming transmitter. Employing several of these systems with the desired phase difference through a phase-locked loop (PLL) enables a very high resolution programmable beam-forming transmitter. Section 4.3 summarizes the chapter.

4.1 Force Mode Distributed Wave Oscillation Mechanisms

The force mode distributed wave oscillator (FMDWO) technique described in this work proposes an alternative approach for a wave oscillator that can

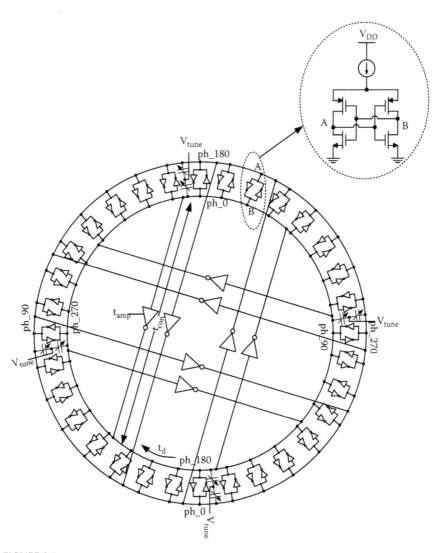

FIGURE 4.1
Delay-based FMDWO circuit diagram.

provide unique advantages, such as even phase symmetry, compared to existing methods. The technique involves additional force mechanisms that create oscillation in two independent differential conductors. The schematic diagram for the delay-based first force mechanism is shown in Figure 4.1. The circuit is built around the two independent transmission line loops. The cross-coupled distributed inverting amplifiers shown in the figure boost the differential traveling wave and compensate for the losses along the line, helping to sustain the oscillation. The other set of inverting amplifiers boost the signal at particular injection points and apply it to the opposite phase points along the same line. In order to obtain an efficient and effective injection, the time delay of the inverting amplifiers and the connecting lines need to be accommodated by delaying the injection point along the lines in an amount corresponding to the sum of these nonideal delays. This would guarantee the inverted injection phase to match the traveling wave phase. The application of this compensating offset delay also determines the wave propagation direction since the wave traveling toward the injection point becomes in phase with the injection signal. The delay equation relating to the components shown in Figure 4.1 can be written as $t_d = t_{amp} + t_{con}$, where t_d is the delay along the traveling wave line, t_{amp} is the inverting amplifier delay, and t_{con} is the connecting line delay. Such an additional mechanism forcing opposite polarity signal across the same conductor with inverting amplifiers prevents the latch-up case and forces a traveling wave along these differential lines. The varactors shown in the schematic are used to tune the frequency to use the oscillator in PLL configuration as well as to modulate the oscillator with baseband data.

In the symmetry-based second force mechanism shown in Figure 4.2, the loop geometry is arranged to bring 2^N injection points into the center of the structure ($2^2 = 4$ point symmetry ring structure is shown in Figure 4.2). The forcing inverting amplifiers are placed across these center meeting points, eliminating the delays associated with the lines connecting the opposite phase points. Although this structure eliminates the connecting line delays from the equation, the number of exact equal-distance accurate phase tap points is limited to 2^N. The crossing of differential lines is done at multiple points to ensure the symmetry between the inner and the outer conductor. Again, the varactors are used for frequency tuning.

Figure 4.3 shows a discrete test system that was built on a printed circuit board (PCB) to verify the idea. All the distributed inverters are discrete inverter pairs, and the size of the board is around 25×25 cm to slow down the operating frequency for these inverters to catch up. The measured clock phases of the clocks along the transmission lines are shown in Figure 4.4. The frequency of oscillation was 129 MHz for this particular discrete implementation.

4.2 Single-Ended Force Mode Structures

The single-ended version of the FMDWO is shown in Figure 4.5. In the case of the fully differential system mentioned above, opposite polarity current flowing in each of the differential pair lines will cancel out to a first order, and the fields will be confined locally in this polarized differential structure. Due to the unidirectional current flow in the single-ended case, the structure will introduce radiation. This way, the oscillator structure can be designed to be a radiating antenna element. This new antenna structure is called force

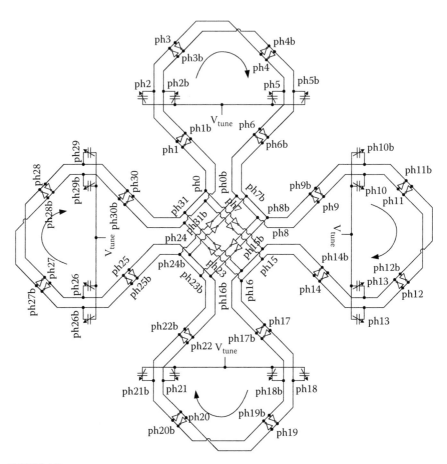

FIGURE 4.2
Symmetry-based FMDWO circuit diagram.

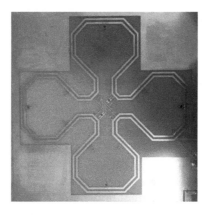

FIGURE 4.3
A discrete implementation of the symmetry-based FMDWO on a PCB.

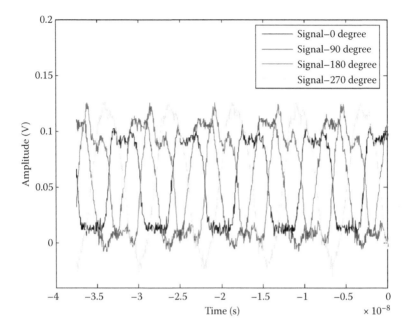

FIGURE 4.4

The measured quadrature traveling wave phases at 129 MHz.

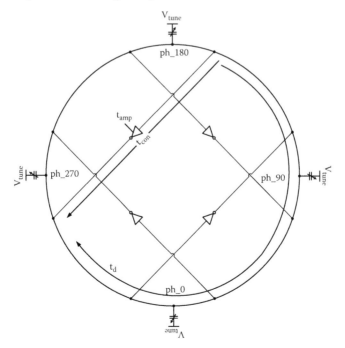

FIGURE 4.5

Single-ended FMDWO forming a radiating element, FMDWA.

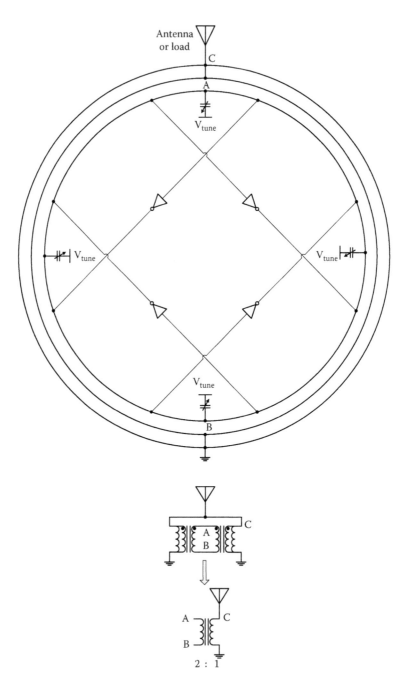

FIGURE 4.6
FMDOA driving antenna loads directly with parallel wave combining configuration.

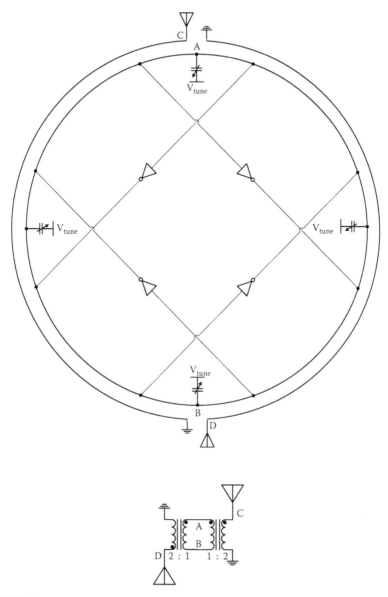

FIGURE 4.7
FMDOA driving antenna loads directly with series wave combining configuration.

mode distributed wave antenna (FMDWA). Using another set of varactors, the frequency of the oscillator can be modulated, resulting in a complete transmitter function. The geometry of the loop structure can be arranged into various closed-loop shapes to obtain an optimal radiation pattern (i.e., closed-loop polygon structures such as square, hexagon, octagon, etc., and meandered path structures).

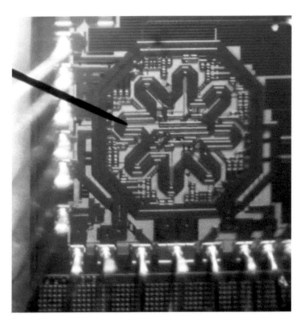

FIGURE 4.8
Die picture of a single-ended FMDWA with secondary pickup coil.

In order to introduce additional degrees of freedom in the antenna choice, a secondary pickup coil is introduced, resulting in a distributed oscillator-transformer combination driving a separate antenna directly. By choosing a secondary side coil configuration, effective voltage or current amplification can be obtained for the antenna drive, as shown in Figures 4.6 and 4.7. Figure 4.6 shows a parallel wave combining architectures that drive different antennas with the corresponding phases. Figure 4.7 shows the series configuration, which has a half turn in the secondary side. This one can also have multiple turns, with an additional half turn to create the opposite phase drives to the two antennas. Having two opposite phase drive pairs of spatially separated antennas also results in beam formation.

A die photo of a test chip that implements this single-ended structure with a secondary pickup coil is shown in Figure 4.8. The IC is implemented in a 0.18 μm complementary metal oxide semiconductor (CMOS) process with six metal layers. The measured signal on the secondary pickup coil driving the spectrum analyzer directly was around −25 dBm at around 10 GHz of oscillation frequency. This power level was significantly less than the expected, due mainly to the losses associated with the bond wire, package, and test board parasitics.

A step-up wave transformation ratio can also be obtained by allowing many series turns in the secondary coil using a symmetry-based approach and multiple crossings, as shown in Figures 4.9 and 4.10. Crossings are the key effect

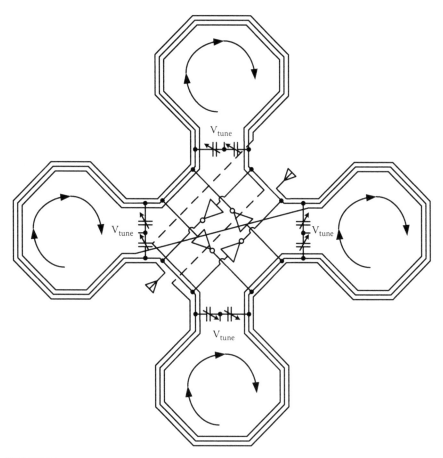

FIGURE 4.9
Dual-secondary-loop FMDOA driving antenna loads directly: two half-loop structures with two antennas each.

for the voltage amplification since they add the waves on top of each other by forcing the maximum amplitude point of a half structure to the minimum phase point of the other half. Figure 4.9 shows a structure with two symmetric halves; each includes a two-turn secondary side with a crossing (shown as dotted lines) connecting the end points. In order to implement a crossing, another metal layer (represented by dotted lines) is needed. The final end points in each half drive two opposite phase antennas. One side can be grounded if a single-antenna system is desired. However, in this case, the impedance terminations at the secondary side will not be symmetric. Figure 4.10 shows two full-cycle secondary side rings with the corresponding crossings at the symmetry center of the structure. In this configuration two antennas are connected to the opposite phase end points of the secondary ring. More turns in this secondary loop will result in more voltage amplification. The structure

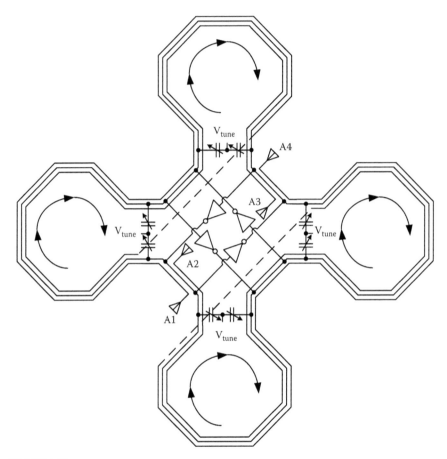

FIGURE 4.10
Dual-secondary-loop FMDOA driving antenna loads directly: one full-loop structure with two antennas total.

effectively forces the waves on top of each other, and hence creates larger signal amplitudes.

The half-symmetry structure of Figure 4.9 and the full-cycle structure of Figure 4.10 can be combined as shown in Figure 4.11, resulting in another configuration. This combined structure connects two symmetric multiple-turn halves of the secondary ring with a crossing (dotted line) at the center, resulting in a similar wave-combining voltage step-up transformer. Unlike classical power amplification techniques, the matching and transformation networks are avoided thanks to the direct antenna drive wave combiner. These wave-combining amplification techniques described are named force mode distributed oscillator amplifiers (FMDOAs).

In high frequency communication systems, the radiation angle from the antenna becomes very narrow, necessitating a programmable beam forming to

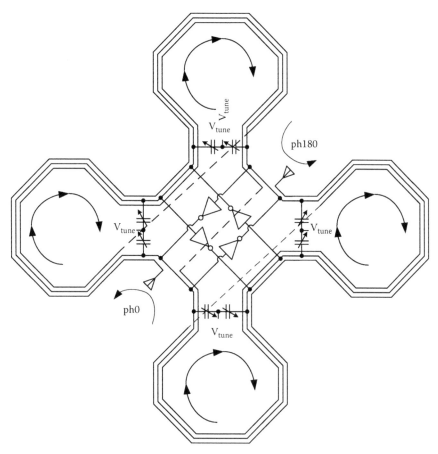

FIGURE 4.11
Two symmetric secondary coils combined into a single two-phase structure with center cross-connection.

establish a directed link based on the location of the transmitter and receiver. This is most commonly done by driving spatially separated multiple antennas with different phases of the signal (phased-array antennas). By changing the phases driving the antennas, a constructive spatial interference of the waves is also changed, resulting in stronger wave propagation in the corresponding direction. In order to create such programmability in the beam direction, multiple FMDOAs can be used, each of them being locked to a slightly different phase of a reference frequency source using a phase-locked loop (Figure 4.12). A programmable delay element in individual FMDOAs sets the final phase for each of the corresponding outputs. By changing the delay in each element independently with a fine resolution, one can obtain fine resolution in the final beam direction. Figure 4.12 shows four dual-loop FMDOAs effectively

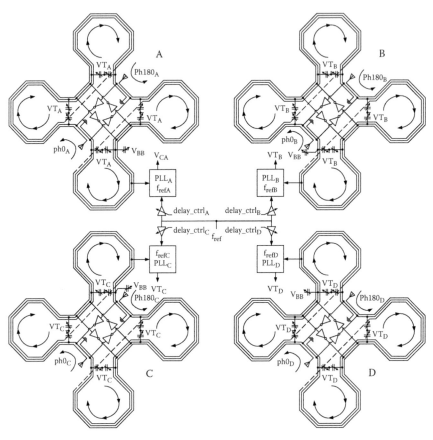

FIGURE 4.12
Four dual-loop FMDOAs driving an eight-element antenna array directly.

forming an eight-element antenna array driver. All the elements are modulated by the same baseband data, V_{BB}, serving as a complete transmitter function.

4.3 Conclusion

A new class of dual-loop traveling wave oscillators has been introduced. Multiple coupling mechanisms have been discussed that can assure differential phase relation on the two independent oscillation tracks for the traveling wave. The same force mechanisms are applied to a single loop to create a traveling wave in a single conductor as a radiating antenna element. Coupling the AC currents to a secondary coil with multiple turns results in transformer action, and hence voltage amplification. Feeding these pickup coils into multiple antennas, various phased-array transmitter structures are reached. The first proof-of-concept test results of some of these ideas were also presented.

Bibliography

1. A. Emira, A. Tekin, D. Ismailov, and A. Suat. *Force-mode distributed wave oscillator and amplifier systems*, US Patent 8,791,765, July 2014.

5

Wave-Based RF Circuit Techniques

As the demand for higher communication bandwidths increases, higher-performance electronic circuit design techniques are explored. Some wave-based radio frequency (RF) circuit techniques are introduced to increase the received signal sensitivity. First, a new transmission line-based traveling wave oscillator technique, quarter-pumped distributed wave oscillator (QPDWO) [3], as a high-purity accurate signal source with multiple oscillation phases is introduced. These high-accuracy, high-frequency oscillation phases open paths to high-performance phased-array transceiver design. Additional noise-cancelling, noise-shaping circuit techniques result in enhanced sensitivity in radio design. Many of the high-frequency electronic oscillator circuits use transmission line-based techniques such as distributed wave oscillators, standing wave oscillators, and rotary traveling wave oscillators that may suffer from drawbacks such as amplitude and phase mismatches, termination impedance mismatches, and added symmetry. Even the higher frequency-capable independent-loop even-symmetry techniques introduced in the previous chapters, trigger mode distributed wave oscillators (TMDWOs) and force mode distributed wave oscillators (FMDWOs), suffer from additional trigger mechanisms that limit the highest frequency that can be achieved. In the case of TMDWO, an additional trigger circuitry is required, whereas in the case of FMDWO, the forcing inverters loading the lines effectively reduce the highest achievable frequency. A new quarter-pumped distributed wave oscillator (QPDWO) technique that avoids these start-up latch-up issues has been described, along with its usage as a phased-array transceiver for a high-sensitivity radio design. At mm-wave frequencies beam forming is a well-known technique to increase the radio sensitivity and hence the range in which it can operate [4–6,8,9]. The phases of the mentioned pumped distributed wave oscillators are used to devise a phased-array front end. Reducing the circuit noise in a receiver is also an important factor in increasing the receiver sensitivity, and hence the radio range. The noise contribution of a receiver can be attributed to two main mechanisms. First is the inherent noise of the receiver circuitry. The noise of the first active receiver block, low-noise amplifier (LNA), is very critical in a radio design. The noise contribution of the following circuitry in the chain is attenuated proportional to the gain in the LNA. The noise cancellation technique described in [7] can limit the LNA noise contribution for a given gain level. Another noise mechanism in a receiver system is the noise resulting from the nonlinear mixing of strong interferers in the vicinity of the desired signal channel. Integrated receivers generally perform channel filtering and variable gain amplification at baseband. The presence of strong adjacent channel blockers along with the desired signal requires a filter with high linearity and dynamic range to attenuate these interferers. The filter must be able to process large signals with little intermodulation distortion. Harmonics of the signal will remain in the filter stop-band where they are automatically attenuated. However, it is very possible that third-order intermodulation between particular combinations of two interfering tones in the stop-band generates significant products in the pass-band. Moreover, since this filter is the first in the chain following the LNA and mixer, its noise contribution remains significant in determining the overall noise figure of the radio, and hence needs to be minimized. Noise-shaping filtering techniques become handy in achieving this goal [1,2]. A frequency-dependent negative resistance (FDNR)-based noise-shaped filter immediately at the mixer output provides a noise-shaped high-order filtering at this node, hence relaxing the linearity of both the following and preceding blocks [2]. Noise of the op-amps and of the passive resistors is high-pass shaped, reducing the total noise in the desired channel. In [1], another second-order current mode filtering technique that results in lower in-band noise is described. The key benefit of low in-band noise filtering at early stages of the receiver chain is to attack the blockers in the early stages of the chain so that the intermodulation noise of the adjacent channels does not limit the sensitivity. In this chapter, these RF techniques are uniquely utilized to propose a new phased-array radio. Section 5.1 presents three pumped oscillator topologies. The low-noise phased-array transceiver with pumped oscillator is covered in Section 5.2, and Section 5.3 is the conclusion.

5.1 Pumped Distributed Wave Oscillators

The quarter-pumped distributed wave oscillator (QPDWO) and the other related RF circuit techniques described in this chapter provide unique advantages to high frequency, high-bandwidth integrated wireless/wireline communication chipsets in the range 0.5 GHz up to 0.5 THz. As was pointed out in detail in Chapters 3 and 4, even the recently introduced high-speed independent-loop dual-traveling-wave oscillators, such as TMDWOs, and FMDWOs, suffer from additional trigger mechanisms that limit the highest oscillation frequency. In this work, two independent transmission line loops form a differential medium for a traveling wave oscillation using noncomplementary (only NMOS(NPN) or only PMOS(PNP)) inverting amplifiers that are biased through a quarter-wave-length transmission line. Use of noncomplementary inverting amplifiers and biasing them through λ/4 transmission lines prevents the latch-up of the oscillation lines. The schematic diagram of QPDWO with NMOS cross-coupled inverting amplifiers is shown in Figure 5.1. The cross-coupled amplifier NMOS devices across the core

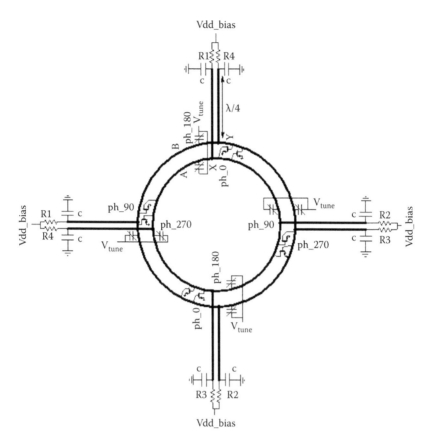

FIGURE 5.1
QPDWO schematic.

oscillator wave tracks A and B are all biased to VDD_osc through $\lambda/4$ shorted transmission line stubs, which effectively poses high impedance at the injection points X and Y and does not load the core oscillator lines. Thanks to the negative transconductance in the inverting cross-coupled amplifiers, a traveling wave oscillation is observed along the two independent transmission lines, each carrying opposite phases of the oscillation at any particular location along these parallel running differential lines. The waves traveling on these two independent loops possess the same propagation characteristic since the lines and the loading corresponding to each line are identical and symmetric. The supply biasing $\lambda/4$ short transmission line stubs and cross-coupled amplifier unit cells should be distributed in maximally symmetric fashion, resulting in a smooth traveling wave. These inverting amplifiers use the signal in one of the lines as a booster for the opposing phase traveling in the other line.

The number of $\lambda/4$ supply stubs can be chosen with respect to the required number of symmetric phase taps. Figure 5.1 shows a four-stub supply distribution scheme. Various techniques can be used to determine the desired wave propagation direction in the core oscillator wave tracks A and B. In one of the techniques shown in Figure 5.1, the supplies to these $\lambda/4$ stubs can be provided through simple *RC* delays, which can implement a mechanism to set the wave direction. Choosing $R_1 > R_2 > R_3 > R_4$ would cause the bias supplies to reach their final values at the corresponding lines, with a proportional amount of delay resulting in one preferred start-up wave direction, whereas the $R_4 > R_3 > R_2 > R_1$ combination results in a wave propagating in the opposite direction. Additional distributed varactors are used to tune the oscillation frequency with a control voltage Vtune, hence enabling its use in phase-locked loops. The shape of the oscillator core tracks, which is shown to be circular in Figure 5.1, can assume any closed-loop shape (square, pentagon, hexagon, octagon, etc.).

A variant of this technique is shown in Figure 5.2. In this case, an inductively pumped distributed wave oscillator (IPDWO) circuit employs inductors (L_b) instead of $\lambda/4$ quarter-wave transmission line to bias the core traveling wave oscillator. The inductors can also be accompanied with parallel capacitors (C_b) to create a resonance at the core traveling wave frequency at the particular bias injection point, resulting in a resonant-pumped distributed wave oscillator (RPDWO). The core traveling wave transmission line loops can be replaced by a lumped structure, as shown in Figure 5.3. Again, the number of bias injection points along the core traveling wave loops, which was four in the figures, can be any number suitable for the desired phase resolution and symmetry.

5.2 Traveling Wave Phased-Array Transceiver

Figure 5.4 shows a phased-array transceiver system utilizing QPDWO, FMDWO, or TMDWO as an accurate source of oscillation phases. These phases directly drive the mixers that are used in the receiver or transmitter path corresponding to each antenna. The figure shows an eight-antenna system that may require an eight-phase oscillator, but the mentioned loop oscillator can be tapped symmetrically to provide a large number of oscillation phases to be selectively mixed with the incoming signal at the front-end mixer. For this front-end phase shift (FEPS) system, all of the phases are routed to all of the mixers to provide programmable phase shift in the corresponding path. Classical phased-array systems need to employ phase interpolators to provide the desired oscillation phases. Phase interpolators consume a significant amount of power and add noise. In the case of these figures, however, by using accurate oscillation phases of TMDWO, FMDWO, or QPDWO, the extra power consumption and noise contribution of interpolators are avoided.

Sending only one set of quadrature oscillation phases to each mixer path and applying a programmable phase shift at the baseband reduces the loading

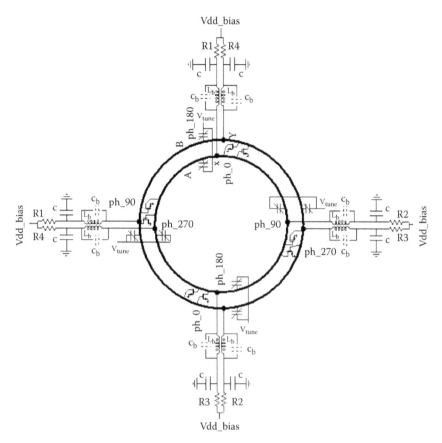

FIGURE 5.2
IPDWO/RPDWO schematic.

to QPDWO, FMDWO, or TMDWO, and hence can result in higher oscillation frequencies. This back-end phase shift system is depicted in Figure 5.5. The down-converted quadrature signals are cross-coupled with proper gain coefficients, resulting in an effective gain and phase shift as follows:

$$A_i I \mp \sqrt{1 - A_i^2} Q + j \left(A_i Q \pm \sqrt{1 - A_i^2} I \right) = \left(A_i \pm j \sqrt{1 - A_i^2} \right) (I + j Q) \quad (5.1)$$

$$phase\ shift = \phi_i = \tan^{-1} \left(\frac{\sqrt{1 - A_i^2}}{A_i} \right) \quad (5.2)$$

For $-1 < A_i < 1$, then $0 < \phi_i < 180°$. Hence, a programmable phase shift at each path in the range $-180°$ up to $180°$ results in another phased-array scheme that can reduce the effective loading on the wave oscillator.

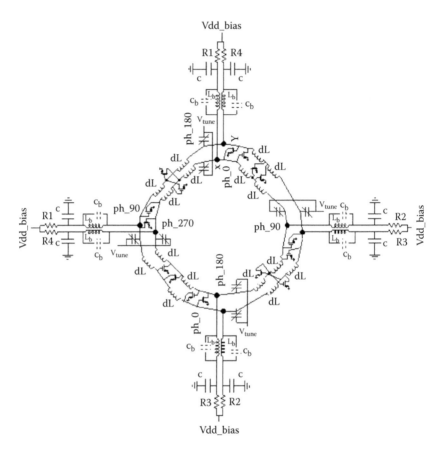

FIGURE 5.3
IPDWO/RPDWO with lumped traveling wave loops.

A schematic diagram of a four-element front-end phase shift system is shown in Figure 5.6. One of the wave oscillators mentioned in this figure, TMDWO, FMDWO, or QPDWO, can be tapped symmetrically to provide the required oscillation phases. If desired, arrays with many more elements can be constructed tapping more phases from these oscillator structures. The LNAs in the received signal paths are followed by mixers that down-convert the high frequency content with the desired phase shift through a programmable phase selector. The corresponding quadrature outputs of all signal paths are then combined and filtered if needed.

The circuit technique introduced in this figure provides balun, LNA, mixer, combiner, and noise-shaped filter functions all in one folded circuit stage, as shown in Figure 5.7. In addition to signal quality enhancement by the phased-array structure, using a minimal number of transistors in the signal paths results in a low-noise, high-sensitivity receiver architecture. These circuit techniques can also be utilized in single-path nonarray receivers. In this

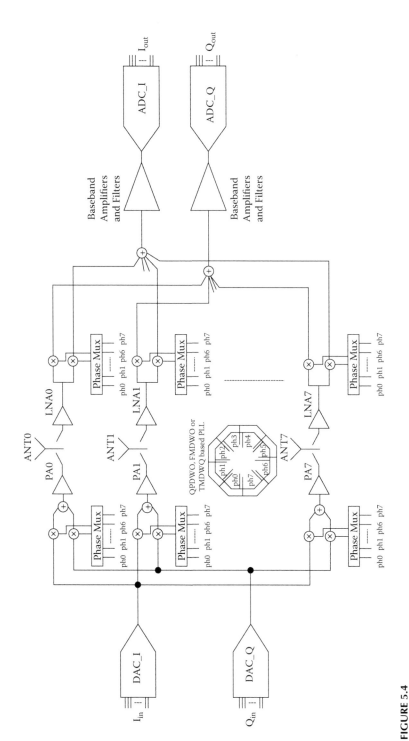

FIGURE 5.4
Wave-based front-end phased-array transceiver system.

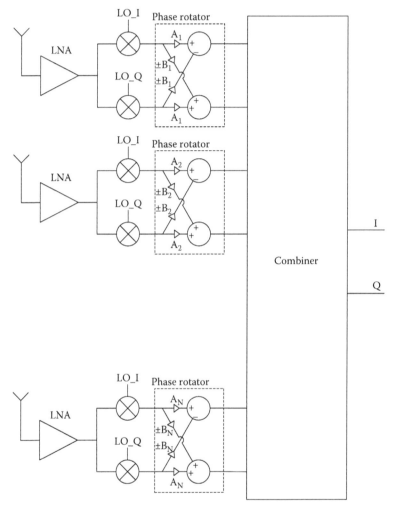

FIGURE 5.5
Wave-based back-end phased-array system.

topology, which is referred to as Blixelter from now on, the outputs of all of the front-end quadrature LNA-mixer pairs (four in this example), each receiving signal from the corresponding antenna element, are combined at the low-impedance summing junctions $sj I_p - sj I_n, sj Q_p - sj Q_n$. RF input drives the IQ noise canceling LNA device pairs $M_0 - M_1$ and $M_{10} - M_{11}$. These common gate, common source amplifier pairs also provide balun functionality, converting the single-ended signal to differential. Quadrature oscillator phases *ph*0, *ph*90, *ph*180, and *ph*270 mix the RF signal down through mixer devices $M_2 - M_5$ and $M_{12} - M_{15}$. The programmable phase mux devices (M_{mux}) in

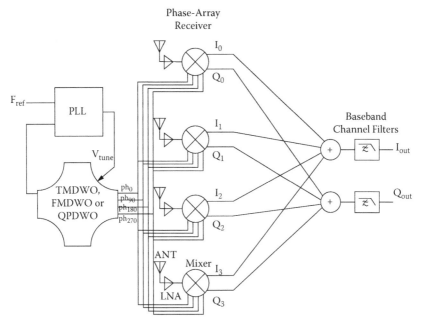

FIGURE 5.6
Wave-based eight-element phased-array receiver block diagram.

each of the signal paths direct the desired IQ phase combination to the combiner summing junctions $sjI_p - sjI_n, sjQ_p - sjQ_n$. Four current sources I_{b1} through I_{b4} provide DC bias current for the four front-end LNA-mixer pairs and some remaining current to bias the combiner. The combiner implements a second-order pipe filter through C_{f1} and C_{f2} and additional third-order notch filtering at the load through FDNR and C_x. Since both of the techniques provide noise-shaped filtering, the strong interferers at the adjacent channels are attenuated without significant addition of filter noise. The receiver front ends implementing such topology can allow higher gain without suffering from adjacent interferers, and hence can achieve better sensitivity. The detailed schematic of the FDNR circuit is shown in Figure 5.8. The overall filter transfer function for the combiner section can be written as

$$\frac{V_{out}(s)}{I_{in}(s)} = \frac{\left(\frac{g_m^2}{C_{f1}C_{f2}}\right) R_f \left(s^2 D R_z + 1\right)}{\left(s^3 D R_z R_L C_x + s^2 \left(D R_z + D R_L\right) + s R_L C_x + 1\right) \left(s^2 + s\frac{g_m}{C_{f1}} + \frac{g_m^2}{C_{f1}C_{f2}}\right)}$$

$$(5.3)$$

where $D = C_1 C_2 R_1 R_3 / R_2$ of FDNR and g_m are the transconductance for the combiner devices $M_6 - M_9$ and $M_{16} - M_{19}$. The filter capacitors C_x and C_2, which are shown to be single ended conceptually in the schematics, can be implemented differentially.

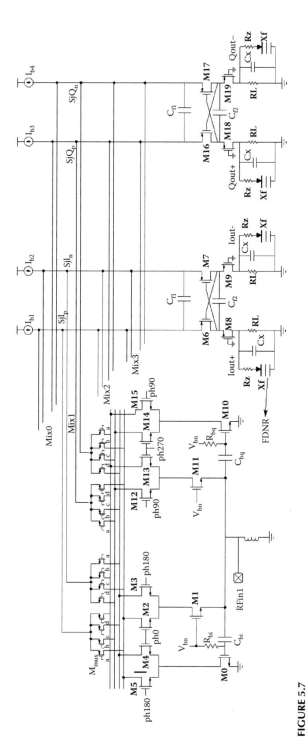

FIGURE 5.7
Detailed circuit schematic of wave-based four-element phased-array receiver.

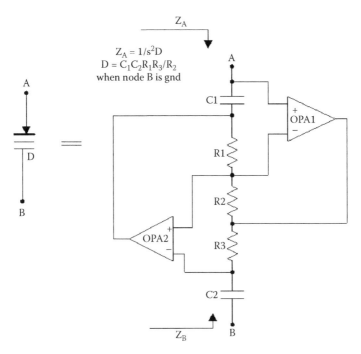

FIGURE 5.8
FDNR circuit schematic.

5.3 Conclusion

New multiphase distributed microwave oscillator topology along with additional noise-canceling, noise-shaping techniques resulted in low-power, high-performance RF transceiver topology. Thanks to the multiphase oscillator, the phased-array front end became a relatively easier task. Since the mixer loads are all absorbed into the oscillator tank, power-hungry LO buffers were all avoided. Many of the radio blocks, such as balun, LNA, mixer, and filters, have all been combined into a single folded stage, resulting in a significant amount of power savings.

Bibliography

1. A. Pirola et al. Current-mode, wcdma channel filter with inband noise shaping. *IEEE Journal Solid-State Circuits*, 45(9):1770–1780, Sept. 2010.
2. A. Tekin et al. Noise-shaping gain-filtering techniques for integrated receivers. *IEEE Journal Solid-State Circuits*, 44(10):2689–2701, Oct. 2009.

3. A. Tekin and A. Emira. Pumped distributed wave oscillator system, US Patent pending 20130157584A1, June 2013.
4. D. Kalian, and J. R. Felland. Phased array antenna systems and methods, US Patent 7,545,324, June 2009.
5. W.E. McCune, Jr. Phased array receivers and methods employing phase shifting downconverters, US Patent 7,859,459, Dec. 2010.
6. H. Hashemi et al. A fully integrated 24ghz 8-path phased-array receiver in silicon. In *IEEE International Solid-State Circuits Conference*, Feb. 2004.
7. S. C. Blaakmeer et al. The blixer, a wideband balun-lna-iq-mixer topology. *IEEE Journal of Solid State Circuits*, 43(12):2706–2715null, Dec. 2008.
8. S. Kobayakawa, Y. Tanaka, and M. Tsutsui. Array antenna receiving device, US Patent 6,208,294, March 2001.
9. D. L. Schilling. Phased array spread sprectrum receiver, US Patent 6400756 B2, June 2002.

6

THz Signal Generation and Sensing Techniques

In this chapter, electronic circuit techniques for THz-frequency signal generation and utilization are described. THz-range high frequency electrical signal generation is known to be a difficult task [1–10]. A new approach uniquely utilizing the accurate symmetric phases of dual-loop traveling wave oscillators such as TMDWO, FMDWO, and PDWO for harmonic frequency extraction is introduced. Since the number of phases available for high-order harmonic boost is decoupled from the maximum achievable fundamental oscillation frequency, the technique provides opportunity for an order of magnitude higher oscillation frequencies than the existing techniques. The technique leads to a unique THz transceiver system that can generate, transmit, and sense THz frequency signals effectively. A method of creating an array of the mentioned single-transceiver sensor element across a semiconductor wafer is disclosed, resulting in an adaptive THz sensing and imaging device for multipurpose use. An adaptive scaling of the programmable system parameters with various calibration surfaces corresponding to a specific application results in a single programmable sensing/imaging device.

Terahertz radiation falls in between infrared radiation and microwave radiation in the electromagnetic spectrum. Similar to infrared and microwave radiation, terahertz radiation also travels in a line of sight. Like microwave radiation, terahertz radiation can penetrate a wide variety of nonconducting materials. Terahertz radiation can pass through clothing, paper, cardboard, wood, masonry, plastic, and ceramics, but it has limited penetration through fog and clouds and cannot penetrate liquid water or metal, a characteristic similar to microwaves.

Due to absorption characteristics of earth's atmosphere, the range of terahertz radiation is limited enough to make it a poor choice for long-distance communication. However, there may still be indoor applications such as high-bandwidth local wireless networking systems. THz sensing technologies, though, provide unique opportunities for a variety of emerging medical and consumer applications, such as cancer detection, detection of biological and chemical hazardous agents (explosives), security, wideband high-speed communications, and radioastronomy. The frequency range of 0.3 to 3 THz is very useful since many substances show a unique absorption/reflection response at a specific corresponding frequency. Although the applications of the idea are abundant, the enabling technologies have significant limitations in terms of system complexity (size), cost, and noise susceptibility due to high attenuation with low signal power. Most existing systems employ laser-based bulky optical THz signal generation techniques as well as large-scale high-accuracy focusing lenses that may not lend themselves to low-cost products.

The chapter is organized as follows; Section 6.1 presents a traveling wave-based frequency multiplication method. Some reflection sensing microwave front-end topologies are discussed in Section 6.2. Section 6.3 describes the high-level system integration and calibration, and Section 6.4 is the conclusion.

6.1 Frequency Multiplication Techniques

The dual-loop coupled traveling wave oscillators of previous chapters, such as trigger mode distributed wave oscillators (TMDWOs), force mode distributed wave oscillators (FMDWOs), and pumped distributed wave oscillators (PDWOs), are very high frequency multiphase oscillation techniques that can provide multiple accurate oscillation phases. In the case of ring type oscillators, the number of oscillation phases is proportional to the number of delay elements used. The oscillation frequency of such oscillators, on the other hand, is inversely proportional to the oscillation frequency. Hence, multiphase capability and highest achievable frequency are two contradicting constraints in ring-type multiphase oscillators. In the case of transmission line-based traveling wave oscillators, however, these constraints are not contradicting since the theoretically infinite number of oscillation phases are readily available along the traveling wave transmission line tracks. Moreover, these types of oscillators can provide a much higher fundamental oscillation frequency than lumped or ring-type oscillators due to their distributed nature. Figure 6.1 shows the core idea of the THz signal generation technique using PDWO as an example, whereas any of the closed-loop transmission line-based multiphase oscillators can be used since all these topologies differ only in the way they trigger a successful oscillation in the traveling wave lines. Hence, in the figures, the core traveling wave oscillators will only be represented as dual-loop tracks without the associated supporting circuitry to focus on the details of the current work. In Figure 6.1, a PDWO including distributed cross-coupled active amplifying devices M1 and M2, distributed frequency tuning varactors C1 and C2, distributed pump inductors Lp, and the differential traveling wave oscillation tracks is tapped symmetrically at five locations corresponding to the five phases of the traveling wave. These five symmetric phases were combined through a selective matching network MN that drives the desired load. When added together, the first, second, third, and fourth harmonics of the signal cancel out, whereas the fifth harmonics of the traveling

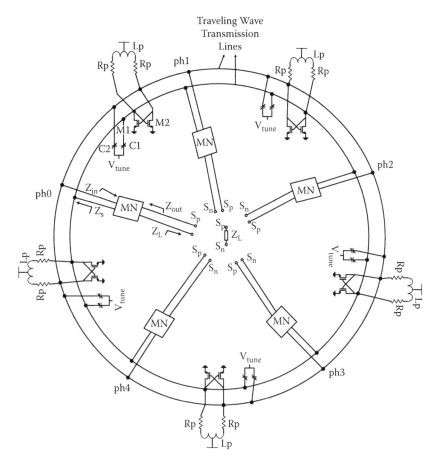

FIGURE 6.1
Traveling wave harmonic extraction technique with PDWO.

wave signals all add up since they would all be in phase. The equations below illustrate this case.

The fundamental frequency content at locations *phi* (for $i = 0, 1, \ldots, 5$) is

$$ph0(t) = A\cos(\omega t + 0 \times 2\pi/5) \tag{6.1}$$

$$ph1(t) = A\cos(\omega t + 1 \times 2\pi/5) \tag{6.2}$$

$$ph2(t) = A\cos(\omega t + 2 \times 2\pi/5) \tag{6.3}$$

$$ph3(t) = A\cos(\omega t + 3 \times 2\pi/5) \tag{6.4}$$

$$ph4(t) = A\cos(\omega t + 4 \times 2\pi/5) \tag{6.5}$$

Since the five equal-amplitude signals with equally spaced phases add up to zero, combining the signals at these five symmetric tap points along the oscillation loop will ideally have no fundamental frequency content. The second, third, and fourth harmonics contents of the signal bear similar phase relationships, resulting in complete cancelation at the center summing nodes, Sp, and Sn. The fifth-order harmonic contents at these tap locations, however, would all be in phase and add up at the summing nodes Sp and Sn. The equations below illustrate this phase relation for the fifth-order signal content. The fifth-order harmonic content at location *phi* (for $i = 0, 1, \ldots, 5$) is

$$ph0(t) = A\cos\left(5\left(\omega t + 0 \times 2\pi/5\right)\right) = A\cos\left(5\omega t\right) \tag{6.6}$$

$$ph1(t) = A\cos\left(5\left(\omega t + 1 \times 2\pi/5\right)\right) = A\cos\left(5\omega t + 2\pi\right) \tag{6.7}$$

$$ph2(t) = A\cos\left(5\left(\omega t + 2 \times 2\pi/5\right)\right) = A\cos\left(5\omega t + 4\pi\right) \tag{6.8}$$

$$ph3(t) = A\cos\left(5\left(\omega t + 3 \times 2\pi/5\right)\right) = A\cos\left(5\omega t + 6\pi\right) \tag{6.9}$$

$$ph4(t) = A\cos\left(5\left(\omega t + 4 \times 2\pi/5\right)\right) = A\cos\left(5\omega t + 8\pi\right) \tag{6.10}$$

This method, which shows frequency multiplication by 5 in this example, can be implemented for a much bigger multiplication ratio N, tapping the mentioned traveling wave arts at N symmetric points. For example, one can ideally tap a 100 GHz one at 15 symmetric points and drain 15th-order harmonic content through 15 symmetric paths as an effective way to generate a 1.5 THz signal. The combining is made more efficient by introducing a selective matching network along the harmonic drainage paths, which displays high impedance at the fundamental frequency not to load the oscillator, whereas it matches the oscillator source impedance Zs to the load impedance ZL at the desired harmonic frequency to be able to transfer maximum harmonic content to the load.

A schematic diagram of the circuit achieving this selective power combining is shown in Figure 6.2. The matching element in the power combining network is only a quarter-wave transmission line of the fundamental frequency ($\lambda/4$), which is also equivalent to a quarter-wave length for the fifth-order harmonic signal, as shown below:

$$\frac{\lambda}{4} - \lambda_{5th} = \frac{\lambda}{4} - \frac{\lambda}{5} = \frac{\lambda}{20} = \frac{\lambda_{5th}}{4} \tag{6.11}$$

Generalizing this relation to any odd harmonic content with order $2k + 1$,

$$\frac{\lambda}{4} - \lambda_{(2k+1)th} = \frac{\lambda}{4} - \frac{\lambda}{2k+1} = \frac{\lambda\left(2k - 4 + 1\right)}{4\left(2k + 1\right)} = \frac{\left(2k - 4 + 1\right)\lambda_{(2k+1)th}}{4} \tag{6.12}$$

From the above equations, one can conclude that a quarter-wave length at fundamental frequency does also correspond to a quarter-wave equivalent of all of the odd order harmonics. Referring back to the power combining scheme of Figure 6.2, since the signal summing points are virtual ground for the

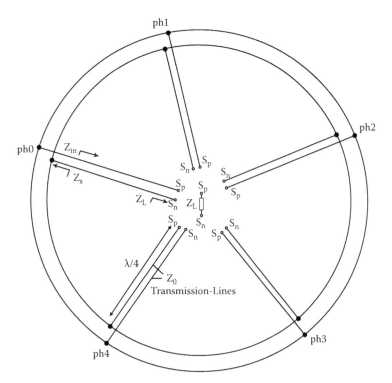

FIGURE 6.2
$\lambda/4$ transmission line-based harmonic matching and combining network in TWFMs.

fundamental frequency signal, the quarter-wave length stub will transform this virtually zero impedance to ideally infinite impedance at the tap points ($Z_{in} = \infty$ at fundamental frequency). Hence, ideally this harmonic drainage network will not load the oscillator at its fundamental operating frequency. With respect to the desired harmonic frequency content, however, choosing the corresponding characteristic impedance Z_0 for the combiner stub, the quarter-wave length can transform a given load impedance Z_L to a Z_{in} value, matching to the source impedance of the oscillator Z_s.

$$Z_{in} = Z_s = \frac{Z_0^2}{Z_L} \tag{6.13}$$

As a result, the technique provides a method that does cancel the fundamental frequency out without loading the oscillator at this frequency, whereas it transfers the maximum power to the load at the desired harmonic frequency. This new technique is named traveling wave frequency multiplier (TWFM).

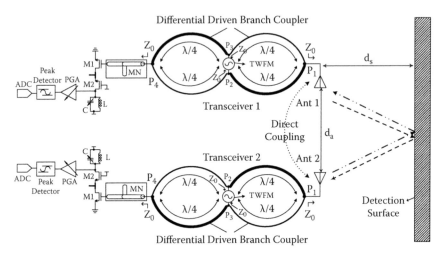

FIGURE 6.3
TWFM-based traveling wave frequency shift reflectometer (TWFSR) transceiver element.

6.2 Traveling Wave Reflectometers

This section goes through two types of reflectometers that target to sense the reflection characteristics of the material to be sensed. In the first system, two identical TWFMs, with a small programmable offset frequency corresponding to desired intermediate frequency (IF), are configured in a unique sensing configuration shown in Figure 6.3. There are two identical transceivers, except a slight frequency offset at the TWFM, that transmit and receive simultaneously. As an example, say if Transceiver1 TWFM is set to generate a 300 GHz tone, then Transceiver2 TWFM can be tuned to 297 or 303 GHz, resulting in an IF of 3 GHz in this example. This IF design parameter can be chosen appropriately for a given technology, interference profile of the environment, and application. The two TWFMs can be tuned to the desired offset frequency very accurately during production programming. Since these devices are identical, their frequency drift behavior with environmental factors, such as temperature, would track each other to a first order, keeping the IF frequency relatively constant.

The mentioned frequency-shifted TWFMs drive differential drive branch couplers (DDBCs), which enable simultaneous reception and transmission in each of the transceivers. This coupler is a special form of a standard four-port branch coupler, where the two symmetric ports are constructed to be physically close so that they can be driven differentially. As in the case of standard couplers, all the ports in DDBCs should be matched and the splitting line impedances should be set accordingly. The DDBC in Transceiver1 directs the half of the received power at its port P1 (303 GHz in the example) to the P4, matched input port of the mixing amplifier. Half of the TWFM power at port P3 (300 GHz) flows into mixing amplifier port P4, whereas the other half travels to P2 with 180° phase shift, becoming in phase with the opposite phase oscillation signal at the same port. Similarly, the power of the opposite TWFM phase at P2 also splits into two, one reaching to the matched antenna as the transmit signal, while the other half propagates to P3, becoming in phase when reaching this port. Thanks to this coupling technique, the TWFM power is split efficiently to be input in the receive path mixing amplifier and at the same time transmit signal to the antenna without disturbing the differential symmetry of TWFM. The received signal, on the other hand, finds a path to the mixing amplifier ideally with only 3 dB loss. So, in respect to the example design parameters presented, Transceiver1 transmits a 300 GHz signal and receives a 303 GHz signal, while Transmitter2 transmits a 303 GHz signal and receives 300 GHz one, reflected from the detection surface. As a result, there exist 300 and 303 GHz signals at the inputs of both mixing amplifiers. This is called traveling wave frequency shift reflectometer (TWFSR) transceiver, which forms a base unit element for the THz sensing/imaging device proposed in the figure. One key factor regarding the performance of the presented art is the direct coupling path from one antenna to the other. Any signal coupling in this path would effectively reduce the system dynamic range if the amplitude of this coupling is comparable with the signal received through the reflective path. In order to mitigate this effect and enhance the system performance, the gap between the antennas da by design is arranged to minimize the antenna-to-antenna coupling. Moreover, the antenna radiation angle is also tuned to yield maximum reflected signal amplitude for a given distance to measurement surface ds. The back ends of the transceivers in this unit element are also identical. The mixing amplifier element in this example is a tuned metal oxide semiconductor (MOS) amplifier, which can be replaced by any active/passive amplifying and mixing circuit topology. In this particular design, the MOS device M1 in the saturation region can very effectively mix down and amplify 300 and 303 GHz down to 3 GHz, due to strong square low characteristics at this operating region. The cascode device M2 improves the reverse isolation as well as the gain of this stage. Following this low-noise mixing amplifying stage, the signal is passed through a chain of programmable gain amplifiers (PGAs) to boost the received tone level to full range of the analog-to-digital converter (ADC). This is followed by a peak detector output that is digitized through an ADC. The digital data represent the amount of reflection through the sensing surface. The gain in the PGA stage can be adjusted to yield the best dynamic range for a particular sensing surface.

The physical layout of the 280 GHz reflectometer is shown in Figure 6.4. The structure includes all biasing single-turn inductors, a differential PDWO, $\lambda/4$ phase combining stubs, branch couplers in the center, and an M1 ground plane in the model. An ultra-thick 3 µm copper layer was used for most of the transmission lines (yellow), whereas the crossings were implemented by 1.2 µm aluminum (red).

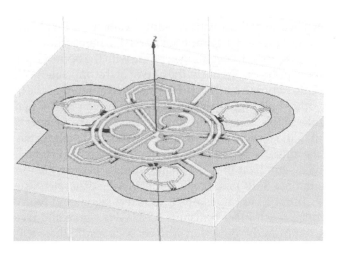

FIGURE 6.4
TWFM-based traveling wave frequency shift reflectometer (TWFSR) transceiver element front-end layout in HFSS.

In the second technique shown in Figure 6.5, a single switching traveling wave frequency multiplier (STWFM) is used as a signal source to symmetrically drive inphase and quadrature (I/Q) signal paths for direct down-conversion of the same frequency-reflected signal reaching each path. The reflected signal that is collected through each of the antennas passes through a symmetrically loaded front-end Wilkinson coupler splitting the power into these two paths. The signal reaching the input ports (P1I and P1Q) of the differentially driven branch couplers finds the path to the input matching network of each corresponding path with 3 dB loss, along with the same frequency STWFM signal into the mixing amplifying device M1. By ON/OFF modulation of the STWFM with a modulation frequency FM, the signal content can be moved to a higher frequency to avoid the impact of DC offset and flicker noise in the final device performance. The down-converted I/Q signals are subsequently amplified, filtered, and finally converted to digital. It is critical to note that the information regarding the measurement surface is embedded in the difference measurement; namely, the system requires a measurement with a calibration surface first, and then with the actual surface to be measured. The deviation from the nominal calibration measurement reflects the characteristics of the surface being measured. The I/Q signals are further processed inside the digital signal processing (DSP) block. In the transmit case, the signals coming from the STWFM through both DDBCs are combined with 90° of phase shift at the front-end Wilkinson coupler feeding the antennas.

The circuit diagram of a STWFM is shown in Figure 6.6. The main difference from a regular TWFM is that the core traveling wave oscillator is powered on and off using a modulating signal FM. When the switching device M3 is turned off, the oscillator outputs settle down to DC bias level Vb. When these devices

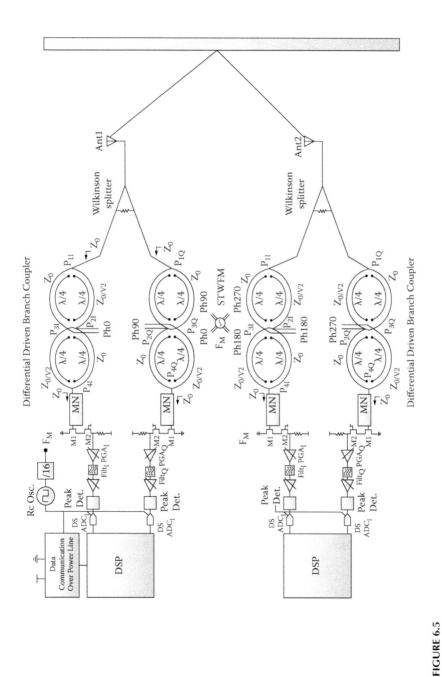

FIGURE 6.5
STWFM-based traveling wave modulation reflectometer (TWMR) transceiver element.

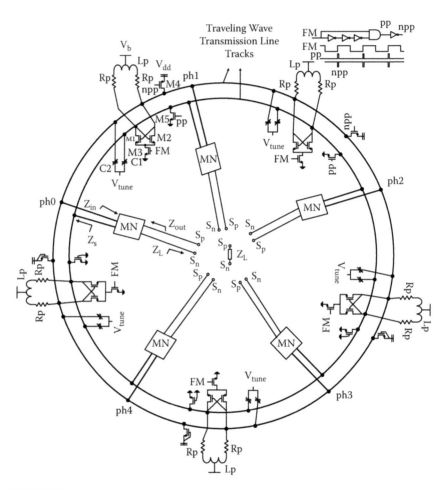

FIGURE 6.6
Switching traveling wave frequency multiplier (STWFM) diagram.

are turned on, the oscillation builds up around the same DC operating bias points. In order to speed up the oscillation buildup, the injection devices M4 and M5 (12 and 13) inject narrow opposite pulses, pp and npp, into the ring. These pulses are generated at the rising edge of the modulating clock FM. A simple well-known pulse generation circuit including delay inverters and an AND gate are used to generate a narrow pulse to drive some of the power-up trigger devices, M4 and M5, on one side of the loop track. Additional dummy devices, M8 and M9, are also added to keep the phases symmetric.

Another topological variation to the presented reflectometer is shown in Figure 6.7. This is based on a dual-traveling-wave oscillator with an offset frequency and referred to as heterodyne traveling wave reflectometer (HTWR). In order to create an offset IF, two traveling oscillators with fundamental

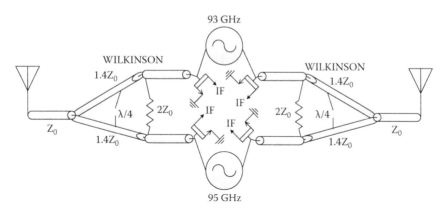

FIGURE 6.7
Heterodyne traveling wave reflectometer (HTWR) front end.

oscillation frequencies ω_a and ω_b are employed. In Figure 6.7, these are 93 and 95 GHz. Each of these signals finds a path to antennas adding together in the Wilkinson coupler, creating the transmit content of both traveling wave oscillators. Hence, the reflected content would contain both frequencies as well. The reflected content splits into the symmetric branches of the Wilkinsin divider and superimposed on top of the existing oscillation signal. For an ideal Wilkinson divider with perfect impedance matching, the only way for one of the frequency contents to appear at the other receiver mixing MOS device is through the reflective path of the antenna. Hence, in order to minimize the direct leakage and reflection paths and increase the sensitivity to the surface to be detected, impedance matching not only around the Wilkinson coupler but also to the antenna needs to be accurate. Moreover, in order for the leakage from the sister oscillator not to pull the frequency with injection, isolation buffers may be inserted following the oscillator.

It is usually almost impossible to create two very close-by oscillations integrated in close vicinity due to pulling effects. However, fully symmetric differential, symmetric traveling wave oscillators come in handy because the fields are confined into the differential traveling wave tracks and inductive coupling is minimal, making it possible to generate them together around the same locality. A possible layout for such a front end is shown in Figure 6.8.

The received signal goes through a MOS device in saturation (square law), which acts as both nonlinear mixer and low-noise amplifier (LNA). The signal content reaching the gate can be written as

$$V_{in} = A\cos(\omega_a t) + \rho[A\cos(\omega_a t + \Phi_a) + A\cos(\omega_b t + \Phi_b)] \qquad (6.14)$$

where ρ represents the reflected wave amplitude. Due to nonlinear devices, in addition to DC, there will be mixing AC IF current content $2a_2 A_\rho(Cos(\omega_a - \omega_b)t - \phi b)$, where a_2 is the second-order term in MOS transfer characteristics.

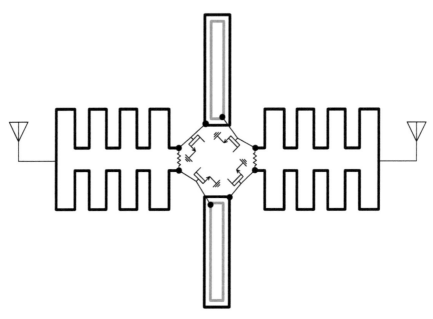

FIGURE 6.8
Heterodyne traveling wave reflectometer (HTWR) front-end layout.

The down-converted IF signal content is pushed to a wideband tuned IF load and amplified through the IF stages. A possible IF stage implementation is shown in Figure 6.9. The RF input components mixing down by the non-linear MOS gain device M1 are pushed through cascode device M2 and reach the tuned load of Lif, Cif, and Rif. Since the frequency difference between the oscillators is not a well-controlled parameter (2 GHz in this example), small value resistor Rif detunes the tank, allowing wideband operation. The signal is then AC coupled to a resistively loaded differential pair programmable

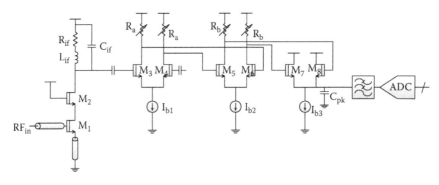

FIGURE 6.9
IF architecture to detect the reflected signals.

gain chain. At the end of the gain chain, a peak detector senses the signal amplitude and delivers it into an ADC following a smoothing filter.

Although larger in area, this heterodyne system may have some unique advantages compared to previously mentioned techniques. First, the low-noise tuned gain in the first stage may result in better sensitivity. Second, although the system can be implemented with a single antenna and two oscillators, adding another antenna and driving them with the opposite phases of these oscillators yields an efficient unit element with double antenna and double oscillator, each transmitting and receiving simultaneously. This way, not only the signal folding back into the same antenna but also the signal originating from the sister antenna and reflecting with angle is collected, resulting in a more idealistic sensing scheme with robustness against surface nonidealities. Moreover, the differential symmetry is preserved by loading the oscillators equally.

6.3 Wafer-Level THz Sensing Method

The TWFSR, TWMR, and HTWR transceiver unit elements that are presented in the previous section can be repeated across a semiconductor wafer resulting in a large-scale near-field THz sensing device, as shown in Figure 6.10. The device dynamic range can be adjusted by optimizing the gain in the system for a specific sensing application. Moreover, in order to enhance the sensing quality and image sensitivity, the device is first calibrated through a calibration surface corresponding to nominal surface characteristics. During this calibration scan of the calibration surface, the gain of each unit element across the array is fine-tuned to yield a uniform clear mid-tone image

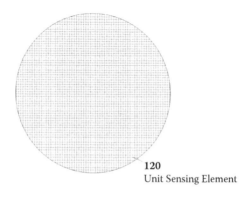

120
Unit Sensing Element

FIGURE 6.10
Wafer-scale TWFSR/TWMR transceiver array for THz sensing/imaging.

corresponding to this uniform calibration surface. This way, any mismatches and nonidealities between the unit elements are all calibrated out, resulting in a uniform and high-accuracy sensing system with high dynamic range. When the real sensing environment is applied, the signal received through each element will very accurately correspond to the surface characteristics of the environment relative to those of the calibration surface. As an example, if the device is to be used in breast or skin cancer detection, first, an artificial calibration surface corresponding to a healthy breast or skin tissue is used to calibrate all of the array elements across the device to fine accuracy. Following this, the device is applied to the patient to get an accurate image of the tissue since any difference from the ideal case will be captured optimally thanks to this calibration technique. The same dynamic range optimization technique can be applied to different calibration surfaces corresponding to different sensing applications. The single device hence can serve multiple THz imaging/sensing applications.

Multiple scanning patterns are implemented to help construct better images based on the raw data obtained from each of the scan results. For example, in addition to one element at a time sensing method to get density, all of the elements surrounding the target unit can also be activated sequentially along with the center element to obtain more information corresponding to this point of interest. Applying a large number of scan patterns and analyzing the data with image construction algorithms, one reaches better image resolution.

6.4 Conclusion

Some unique THz signal generation and utilization methods have been introduced. The main obstacle in silicon-based low-cost THz has been the availability of workable amounts of THz signal power. The traveling wave-based frequency multiplication method was introduced as an alternative way of generating THz signals using silicon-based standard IC production technologies. The techniques discussed herein are expected to open paths to low-cost THz sensing devices for many emerging applications.

Bibliography

1. A. D. Smith and A. H. Silver. Integrated superconductive heterodyne receiver, US Patent 5493719 A, Feb. 1996.
2. B. P. Ginsburg et al. Terahertz phased array system, US Patent 8472884 B2, June 2013.

3. A. Raisanen et al. Capability of schottky-diode multipliers as local oscillators at 1thz. *Microwave and Optical Technology Letters*, 4(1):29–33, 2007.
4. D. Huang, M. F. Chang, and T. R. LaRocca. Submillimeter-wave signal generation by linear superposition of phase-shifted fundamental tone signals, US Patent 8130049 B2, March 2012.
5. P. H. Siegel, R. Dengler, and E. R. Mueller. Sub-millimeter wave frequency heterodyne detection system, US Patent 7,507,963, March 2009.
6. R. S. Cargill and G. E. Mueller, High efficiency low noise frequency tripler and method, US patent 6707344, March 2004.
7. B. Floyd et al. Sige bipolar transceiver circuits operating at 60 ghz. *IEEE Journal of Solid-State Circuits*, 156–167, Jan. 2005.
8. A. I. Grayzel and W. Emswiler. Frequency multiplication by a prime number using multiplier chains. In *IEEE MTT-S International Microwave Symposium Digest*, 321–322, 1978.
9. K. Y. Kang et al. Apparatus and method for generating THz wave by heterodyne optical and electrical waves, US Patent 7684023 B2, March 2010.
10. D. P. Owen. Frequency Multipliers, US Patent 4,400,630, Aug. 1983.

7

Traveling Wave-Based High-Speed Data Conversion Circuits

In this chapter, traveling wave-based data conversion circuit techniques are introduced. Although much of the electronics systems around us are all digital nowadays, the source and destination of these signals in nature remain analog. The signal bandwidths that we are interested in have recently been growing significantly, thanks to the increased demand for high-speed wireless and wireline communications. This in return has boosted the need for high-speed, high-resolution data conversion circuits, namely, analog-to-digital converters (ADCs) and digital-to-analog converters (DACs).

Figure 7.1 shows speed-resolution trade-offs for various ADC topologies. High-resolution and low-bandwidth applications such as audio, video, digital cameras, etc., best utilize sigma-delta-type ADCs. The trade-off comes in the form of signal bandwidth and resolution. For relatively low signal frequencies and a large oversampling ratio, quantization noise of the ADCs can be pushed out of the signal band and filtered, resulting in significantly higher resolution. Digital cameras, toys, appliances, and mobile phones tend to use successive approximation register (SAR)-type ADCs to better fit into the resolution, power consumption, and sampling clock requirements. Emerging 4–5G mobile and other wireless transceiver applications, modems, routers, storage devices, and data acquisition systems may in most cases utilize pipelined ADCs. This type of ADC presents a good trade-off in terms of power consumption and resolution at relatively high input bandwidths. Going up to very high speeds, though, flash or folded architectures become more prominent choices if the application does not pose stringent power and area requirements. These types of ADCs are commonly employed in digital oscilloscopes, high-speed instrumentation equipment, and high-resolution radar systems.

The fast scaling of complementary metal oxide semiconductor (CMOS) process technology has helped a large number of high-speed designs to emerge. More is needed on the design side, however, to compensate for the drawbacks of dynamic range reductions due to shrinking supply voltages. In this respect, Section 7.1 describes a noise-shaping data conversion technique using traveling wave oscillators. Section 7.2 goes into the details of a phase-interleaved ADC architecture with high-speed SERDES as I/O. In Section 7.3, a multiphase high-speed DAC architecture is introduced. Section 7.4 is about beam-forming transmitter DAC, and Section 7.5 is the conclusion.

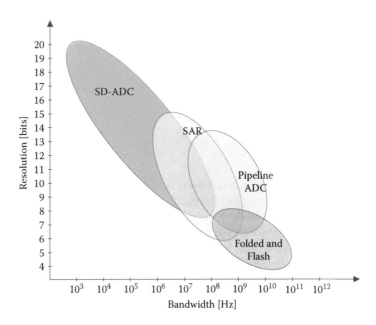

FIGURE 7.1
Speed and resolution trade-off for commonly used ADC topologies.

Figure 7.2 shows a simplified operational block diagram of a SAR ADC. A converging binary search is conducted through a DAC, a comparator, and a logic block resolving N bits in N clock cycles total. The first introduction of such a converter goes back to the 1940s when Bell Labs engineers demonstrated multiple designs based on the successive approximation technique. In 1946, John C. Schelleng of Bell Labs had first introduced his patent in the context of a pulse code communication system [7]. The following year his

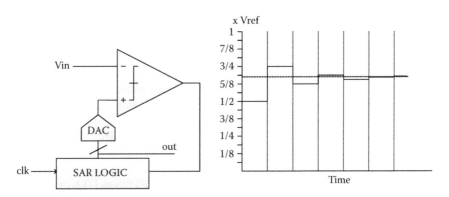

FIGURE 7.2
SAR ADC simplified block diagram.

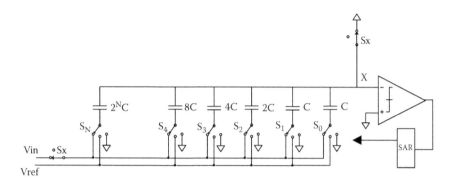

FIGURE 7.3
Capacitive SAR ADC.

colleague W.M. Goodall of Bell Labs introduced 5-bit 8 Kb/s SAR ADC for voice telephony application [4]. The voice signal is first sampled to a capacitor. It is then compared to a reference voltage that is equal to half of the maximum input. If it is greater than this reference voltage, the most significant bit (MSB) is registered as a 1, and an amount of charge equal to half of full scale is subtracted from the storage capacitor. If the voltage on the capacitor is less than half scale, then no charge is removed, and the bit is registered as a 0. After the MSB decision is completed, the cycle continues for the second bit, but with the reference voltage now equal to quarter scale. This technique remains very similar to the capacitive redistribution DAC feedback of today's most commonly used SAR topology. Thanks to the advanced semiconductor fabrication technologies, resolution levels as high as 16 bits can be obtained depending on DAC elements matching. There is a speed limitation due to the fact that N comparison clock cycles are needed before finishing a single conversion.

Figure 7.3 shows one of the most common implementations of SAR topologies, a capacitive DAC implementation. The switches are shown in the track mode where the analog input voltage is constantly charging and discharging the parallel combination of all the capacitors. The hold mode is initiated by opening the sampling switches Sx, leaving the sampled analog input voltage on the capacitor array. Since node X is now floating, the voltage at this node will move as the bit switches are manipulated. If S_1 through S_N are all connected to ground, a voltage equal to $-V_{in}$ is pumped to node X. Connecting S_N to V_{ref} adds a voltage equal to $V_{ref}/2$ to node X. The comparator then makes the MSB bit decision based on the superposition of the voltages at node X ($V_{ref}/2 - V_{in}$), and the SAR either leaves S_N connected to V_{ref} or connects it to ground depending on the comparison result. A similar process is followed for all of the lower significant bits for a total of N comparison cycles. At the end of the conversion interval, the final bit combination corresponding to the input level is latched and the converter starts the next cycle.

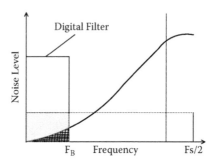

FIGURE 7.4
Oversampling and noise shaping.

Sigma-delta-type ADCs constitute another group of ADCs that have become quite popular with the advance of CMOS process technologies in recent years. The first being introduced in 1960 by C.C. Cutler of Bell Labs [2], the sigma-delta modulation technique has improved significantly. These ADCs trade off resolution with bandwidth and can result in the best signal-to-noise ratio (SNR) numbers at low-bandwidth applications such as audio. In recent years, however, many wireless communication systems have well adapted these types of ADCs as the technology scaling has helped speed to enable oversampling. Oversampling and noise shaping are the key elements of this technique. Oversampling helps SNR by reducing noise falling in-band, as shown in Figure 7.4. Ideally, noise power falling in-band reduces by 3 dB/octave, and hence oversampling yields 0.5 bit/octave resolution improvement. However, employing a noise-shaping loop along with oversampling reduces the in-band noise even further. The assumption here is that the following digital filtering block removes the noise spectrum in excess of the signal bandwidth.

Figure 7.5 shows a general sigma-delta modulator topology. ADC is represented as an ideal converter with quantization noise $E[z]$. $H[z]$ represents the filter transfer function ahead of the transfer function of this loop, which can be expressed as

$$Y[z] = E[z] + H[z]X[z] - H[z]Y[z] \qquad (7.1)$$

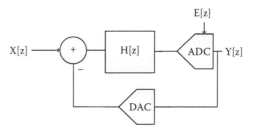

FIGURE 7.5
Simplified sigma-delta modulator diagram.

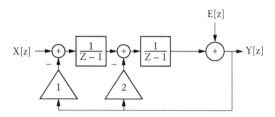

FIGURE 7.6
Second-order sigma-delta modulator loop diagram.

Therefore, we can write

$$Y[z] = \frac{E[z]}{1 + H[z]} + \frac{H[z]}{1 + H[z]} X[z] \qquad (7.2)$$

Utilizing a single integrator for $H[z] = z^{-1}/(1 - z^{-1})$ yields a first-order modulator relating the input and output as

$$Y[z] = E[z](1 - z^{-1}) + z^{-1}X[z] \qquad (7.3)$$

The output presents the signal as well as high-pass-filtered quantization noise. Hence, the noise falling into signal band is reduced.

The concept can be extended to higher-order modulators by utilizing additional integrators in the loop. A diagram showing a second-order loop is shown in Figure 7.6. The input-output relation in this case, including the quantization error, can be written as follows:

$$Y[z] = E[z](1 - z^{-1})^2 + (z^{-1})^2 X[z] \qquad (7.4)$$

The signal transfer function (STF) is unity, whereas the noise transfer function (NTF) has high-pass characteristics with 40 dB/decade, as shown in Figure 7.7. Thus, a higher-order loop effectively yields less in-band noise for a given oversampling ratio. The concept can be expanded to much higher order loops with sophisticated transfer characteristics, but only the basic concept has been discussed in the context of this book to prepare the reader for Section 7.2. In this particular section, the traveling wave-based noise-shaping concept with time domain quantization is introduced.

Figure 7.8 shows flash ADC topology, which is used in high-speed applications. The drawback of the topology is that for every additional bit the power and area approximately double. The number of comparators becomes impractical at the 7-bit level and beyond due to excess power and area requirements.

One remedy for this drawback of flash ADCs is to separate the static preamplifier stages of the comparator and employ interpolation to drive the required number of back-end low-power and slow dynamic comparator latches. A more commonly implemented way around it is folding or subranging. In this topology, a coarse ADC sets the MSBs, whereas a folding circuit generates

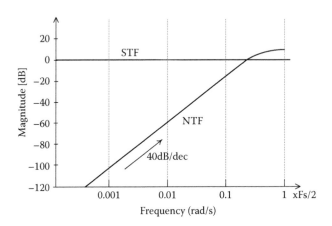

FIGURE 7.7
Second-order sigma-delta modulator transfer characteristics.

a pattern to a parallel flash to extract the LSB simultaneously, as shown in Figure 7.9. Folding linearity is the key in this type of converter. The topology offers high-speed operation with fewer comparators.

Pipeline ADCs also offer high-speed operation by trading latency with speed. This topology was first introduced in 1956 by B.D. Smith [8]. Input

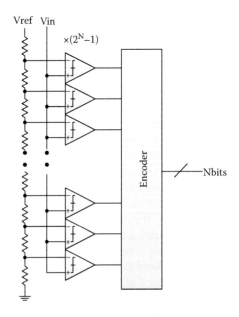

FIGURE 7.8
Flash ADC topology.

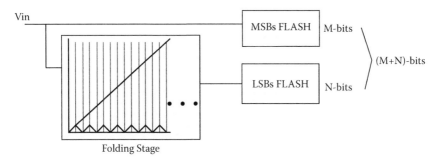

FIGURE 7.9
Folding ADC block diagram.

signal is first sampled and held steady by a sample-and-hold circuit. The system propagates the operation along a chain of stages resolving single or multiple bits along the pipe. While the flash ADC in the first stage quantizes input, output is then fed to an accurate precision DAC, and the analog output is subtracted from the input. The residue signal is then gained up by a factor of two in the case of a single bit in the first stage and fed to the next stage. This gained-up residue continues through the pipeline, until the last flash ADC, which resolves the least significant bits (LSBs). Because the bits from each stage are determined at different points in time, all the bits corresponding to the same sample are aligned with shift registers. Note when a stage finishes processing a sample, determining the bits, and passing the residue to the next stage, it can then start processing the next sample received from the sample-and-hold embedded within each stage. Thus, the pipelining mechanism trades the latency with throughput. An operational block diagram of a three-stage pipeline ADC is shown in Figure 7.10.

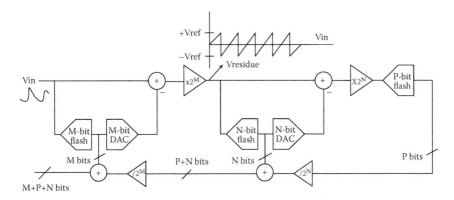

FIGURE 7.10
Pipeline ADC operational diagram.

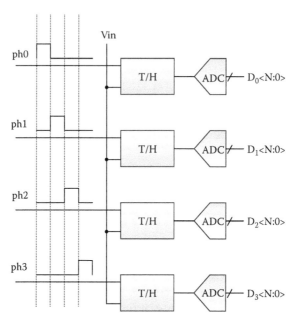

FIGURE 7.11
Time-interleaved ADC system.

Flash, folded, or pipeline ADCs all still have limitations in conversion speed since now they all need to resolve the worst case of LSB signal level for the proposed resolution. Although this is much less pronounced in flash ADCs, it will still be the limiting factor. In order to decouple the conversion rate from the worst case comparator delay, high-speed data converters often employ time interleaving. The idea of a time-interleaved ADC is to use a system of multiple parallel ADCs, which alternately take samples. Thereby, the sampling frequency of one channel does not need to fulfill the Nyquist criterion; however, when in the digital domain, all samples are merged into one output sequence and the overall sampling frequency fulfills the Nyquist criterion. Therefore, sampling with an ideal time-interleaved ADC with N channels is equivalent to sampling with an ideal ADC with an N times higher rate. An example of a four time-interleaved ADC system is shown in Figure 7.11. Each ADC path samples the input at one of four clock phases, converting in the rest of the cycle or in the adjacent full cycle of the clock. Hence, ADC conversion speed is four times lower than the sampling speed, but using four in parallel yields a four times higher-speed conversion rate. In time interleaving systems, the timing mismatch between the clock phases is very critical if sampling the input directly, and hence requires very high accuracy.

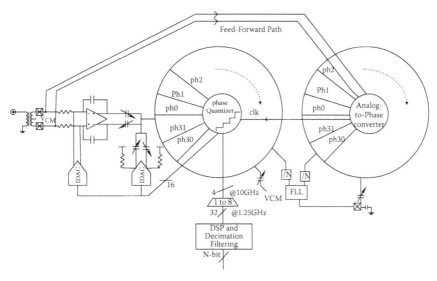

FIGURE 7.12
A traveling wave noise-shaping modulator.

7.1 Traveling Wave Noise-Shaping Modulator

A traveling wave noise-shaping modulator (TWNSM) is shown in Figure 7.12. Many voltage-controlled oscillator (VCO)-based quantizer loops are available in the literature [3,5]. Using traveling wave oscillators, however, can provide a further unique opportunity to design high-speed, high-resolution sigma-delta loops. In the proposed scheme, two identical frequency locked traveling wave oscillators were used. They are represented as circles with 32 symmetric phases in this example for simplicity, but they represent the same multiphase dual-loop traveling wave oscillators: TMDWO, FMDWO, or PDWO. The phases of the one with phase quantizer capture the instant phases corresponding to the sum of the integrator and the feedback path signal. There is an additional integration pole inherent to the VCO, and hence the total modulator loop becomes second order. Thus, when the loops are closed with the feedback DACs, the bit stream captured at the phase quantizer represents the noise-shaped input signal. The 16 phase quanta are fed back to current digital to analog converter (IDACs) and a 4-bit digital signal goes through decimation filtering for *N*-bit final digital output. All the paths are implemented differential, eventually feeding into a differential varactor control into the traveling wave oscillator. The second wave oscillator on the right, the one with analog-to-phase converter, generates the clock that samples the phases in the first wave oscillator. A frequency lock loop (FLL) locks

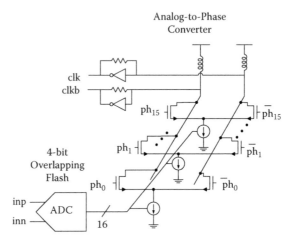

FIGURE 7.13
Analog-to-phase converter.

the phase of this second wave oscillator to the average phase of the first one by filtering the phase difference in the FLL loop. A large value capacitor does this action while the filtered signal is again applied to a varactor V_x bringing the oscillators into lock. The same varactor is added to the first one with a mid-level bias to keep them symmetric. A feed-forward path quantizing the analog inputs into corresponding clk phase (analog-to-phase converter) re-laxes the signal swing at the output of the integrator and relaxes the tuning range of the main wave oscillator with a phase quantizer. Thus, the sampling clock moves in the signal direction and absorbs a large portion of the phase jumps, leaving only a much smaller error signal phase modulation to the main wave oscillator. Again, the most unique advantage of the scheme lies in the availability of high frequency, high-accuracy multiple phases of a wave oscillator clock. In classical methods 32 phases would require a 16-stage dif-ferential ring oscillator, which in return yields quite low oscillation frequency that limits the speed.

The details of the feed-forward analog-to-phase converter, feedback current DAC, and phase quantizer schematics are all shown in Figures 7.13, 7.14, and 7.15, respectively. In the feed-forward path, the output of the 4-bit flash ADC directs the best fitting one of the 16 clock phases, ph_0 to ph_{15}, as the sampling clock to the phase quantizer. The accuracy of this path is not critical, but the result of phase selection should be resolved fully, and hence the sampling clock should propagate to the phase quantizer in the corresponding sampling period. The feedback DACs are current steering differential pairs with 16-unit elements, and the phase quantizer is a simple pass-gate structure that captures the phases of the first oscillator with the sampling clock from the second.

The waveforms depicting the sampling of the phases in the quantizer are shown in Figure 7.16. The rising edge of the sampling clock captures 16 of the

IDAC

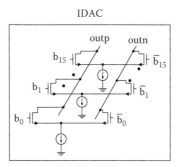

FIGURE 7.14
Sixteen-element feedback DAC.

quantizer clock phases, resulting in a thermometer code that represents the quantized data. It should be noted that half of the phases corresponding to the observation window of the half period are actually used for signal chain processing, while the sampling on the remaining 16 phases is maintained just for the purpose of symmetry.

7.2 A High-Speed Phase Interleaving Topology

Many of today's wireline serial communication links and many radio frequency communication systems require high-speed ADCs. Most such high-speed systems employ multiple relatively low speed ADCs on an interleaved fashion with a high-speed front-end sampler. These front ends require extremely symmetric and low-noise clock phases to sample the incoming high-speed signal. Figure 7.17 shows a traveling wave-based interleaved ADC scheme, which again achieves this target by utilizing the mentioned dual-loop

Phase Quantizer

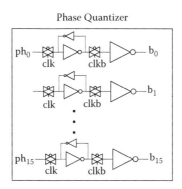

FIGURE 7.15
Sixteen-element phase quantizer.

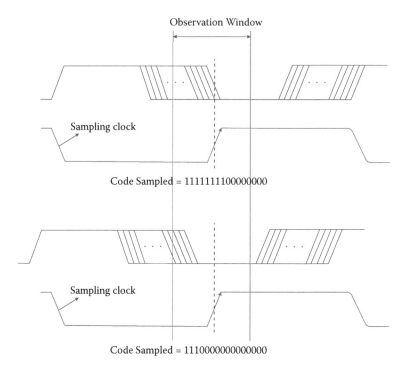

Observation Window

Sampling clock

Code Sampled = 1111111100000000

Sampling clock

Code Sampled = 1110000000000000

FIGURE 7.16
Illustration of the idea of phase sampling and resulting thermometer code.

even-symmetry traveling wave phases. The same clock phases can be used to serialize the parallel output data stream of the ADC for transferring the data to the other chipsets. Again, the high frequency evenly spaced multiple phases are the key feature of the proposed ADC.

Jitter of the signal source is one of the key parameters that affect the performance at high frequencies. In [1], a system for randomizing aperture delay in a time-interleaved ADC system is described. The system may possibly randomize the existing nonidealities, and hence possibly improve the tonal behavior with the expense of higher noise floor. Referencing the sampling clock directly to the oscillator source may yield much desirable jitter performance. The wave oscillator has a tuning mechanism through the distributed varactors to lock it to a low-noise reference source, such as a quartz crystal oscillator, resulting in significantly lower rms jitter. Each of the ADCs shown in the figure can be any appropriate architecture, but the example circuits presented in this work target 8-bit 1-1-1-1-1-3 pipeline topology, with each working at a 3.125 GS/s rate in a 65 nm CMOS technology. All of the required complementary clock phases are tapped directly from the oscillator; hence, all of the 8-bit interleaved pipeline ADC clock load is absorbed into the traveling wave tank. Hence, the power-hungry clock distribution buffers

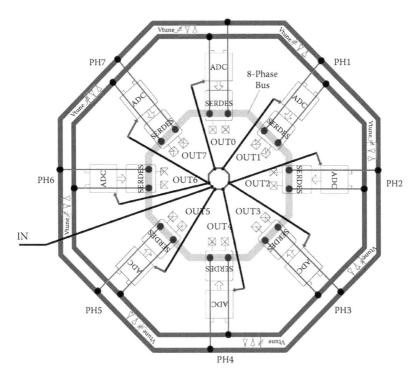

FIGURE 7.17
A traveling wave interleaved ADC with SERDES.

are all completely avoided. The incoming RF signal is terminated to 100 Ω differential with inductive peaking to drive all the ADCs directly without an additional continuous-time or track-and-hold buffer stage. The ADC top-level circuit block diagram is shown in Figure 7.18. Each stage is a single capacitive pipeline stage with a differential comparator and a differential amplifier. The nonoverlapping of the complementary clocks is achieved by adjusting the

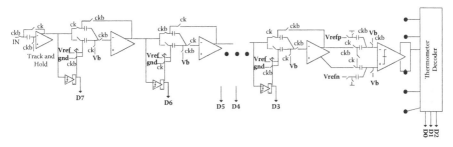

FIGURE 7.18
An 8-bit pipeline ADC schematic.

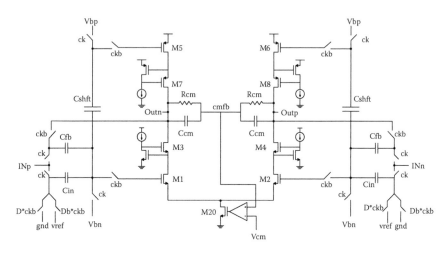

FIGURE 7.19
Amplifier circuit.

pull-up and pull-down strengths, yielding slow turn on while turn off is sharp. The analog inputs are sampled to both the input and the feedback capacitors to maximize the feedback factor in the amplification half cycle and relax the amplifier bandwidth requirements. All of the amplifiers and the comparators are offset compensated during the power-up calibration sequence.

The amplifier is a single gain boosted stage with both PMOS and NMOS differential input pairs. The detailed schematic of the amplifier is shown in Figure 7.19. The input Pdiff pair (M_5-M_6) is biased independently than the Ndiff pair (M_1-M_2) through the capacitors Cshft. Each is DC biased to its respective DC bias points Vbp and Vbn. The p-side is biased to source the desired DC current, whereas a common mode feedback circuit adjusts the n-side current, setting the output common mode for the best signal swing. Both n-side and p-side gain paths employ boosted cascode devices (M_3-M_4 and M_7-M_8) to enhance the low-frequency gain of the amplifier.

The amplifier is designed to work from a 1.8 V supply and can deliver 2.4 $V_{pp-diff}$ signal swing with all 1.2 V thin oxide devices through proper cascode biasing. With a total current consumption of 2.9 mA, it can provide around 10 GHz of unity gain bandwidth and 60 dB DC gain from the mentioned single stage. The simulated unity gain loop characteristics are shown in Figure 7.20. The same network is assumed to be the load for this particular simulation.

Common mode loop characteristics are shown in Figure 7.21. High common mode gain assures accurate common mode levels, and hence implies no sacrifice from the signal swing along the chain.

The comparator circuit that is used in the design is shown in Figure 7.22. It uses a PMOS differential pair preamplifier and a strong-arm latch followed by another cross-coupled NAND latch. The preamplifier is fed from a 1.8 V

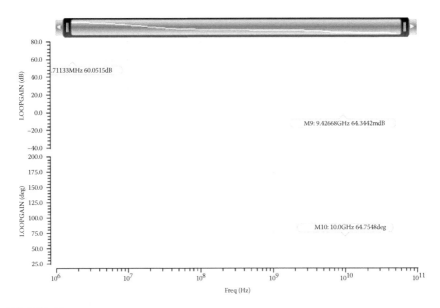

FIGURE 7.20
Amplifier loop characteristics in unity gain configuration.

supply, while the latched comparator uses a 1.2 V rail. Using a PMOS diff pair hence has an advantage of driving the comparator directly without additional protection circuitry. The preamplifier includes DC offset trim resistors Rcal that are set during the power-up calibration sequence. The preamplifier runs with 800 μA of static current.

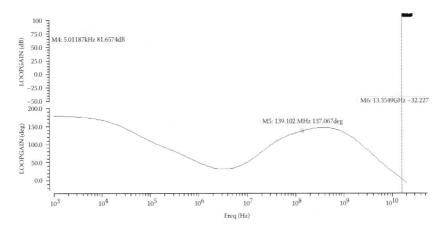

FIGURE 7.21
Amplifier CM loop characteristics in unity gain configuration.

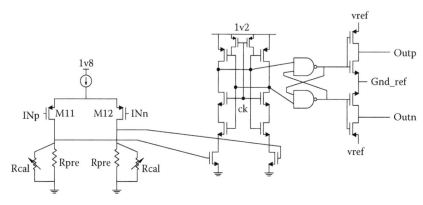

FIGURE 7.22
Latched comparator schematic.

A final 3-bit flash schematic and corresponding timing diagram are shown in Figure 7.23. The 3-bit flash is a capacitive flash with a sampling capacitor value of around 20 fF. Since the LSB for this final stage is around 125 mV, no offset calibration is implemented at this stage. Comparators are plain dynamic latch with no preamplifier stage. In order to allow relaxed timing for

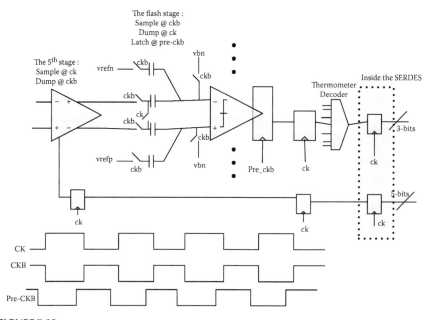

FIGURE 7.23
Final 3-bit flash schematic and timing diagram.

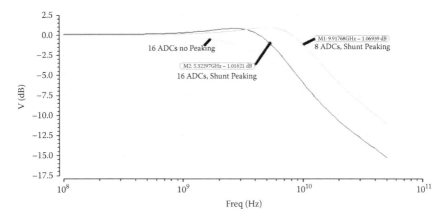

FIGURE 7.24
Input bandwidth with number of ADCs and shunt peaking.

the thermometer decoding, one clock cycle pipelining is implemented. The final alignment latches are included in the SERDES, not to add any further pipelining overhead. Current consumption in the flash is around 600 µA.

A critical design specification in high-speed ADCs is the input bandwidth. As mentioned at the beginning of the section, the architecture avoids any input buffer and the 100 Ω matched transmission line is assumed to be the driver impedance for the system. When loaded with a total of 8 or 16 ADCs, the variation of the input bandwidth with number of ADCs and improvement with on-chip shunt inductive peaking is shown in Figure 7.24.

7.3 A Traveling Wave Multiphase DAC

A traveling wave multiphase interleaved encoded DAC schematic is shown in Figure 7.25. In this sampling system a very high frequency digital-to-analog converter (DAC) method is described. In a classical high-speed DAC, a stream of digital data is sampled by a corresponding high-speed clock and converted to analog. Jitter, edge rate, and other nonidealities associated with a high-speed DAC sampling clock limit the performance in most of the cases. In [6], multiple phases are used to oversample the same DAC input data in order to move the quantization noise to higher frequencies for relatively easier filtering. This does not, however, target higher signal bandwidths in the DAC. In this work, a slower parallel stream of data is sampled into the multiple DACs, with corresponding phases reaching the same very high speed update rate, but with a slow clock. Figure 7.25 shows an example of such a DAC

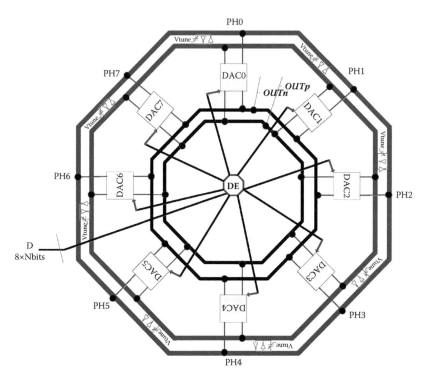

FIGURE 7.25
A traveling-wave multiphase encoded DAC.

architecture with eight phases, which are generated again with any of the symmetric traveling wave oscillators, such as TMDWO, FMDWO, and PDWO. The 8xNbits data stream is latched into the system every traveling wave clock cycle. Inside this multi-DAC summing converter, a simple encoder calculates the code corresponding to each of the eight DACs to yield the desired analog signal level. At every clock phase, the encoded code is applied to the corresponding DAC, resulting in the required incremental change at the output level. Note that although the update rate for each of the DACs is the fundamental clock rate of the traveling wave, the rate of change at the final outputs is eight times faster due to each update corresponding to one of the phases. The encoding is a simple one; for example, to apply the difference between the incoming data to the next DAC, Ddack(k) = Ddata(k) – Ddata(k–1) – Ddack(k–1), where k is the sample number in the sequence of data in the incoming parallel data stream. The DAC corresponding to a clock phase only makes the incremental addition of the level required to make the new sum of all the DAC levels represent the next data sample in the stream. The advantage of the scheme lies in the fact that sampling intervals are defined

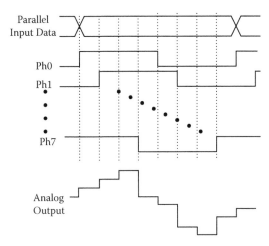

FIGURE 7.26
A traveling wave multiphase encoded DAC waveform.

by the symmetric phases of a rather low frequency clock to reach a very high update rate DAC. Figure 7.26 shows the waveforms corresponding to such a DAC.

7.4 Traveling Wave Phased-Array DAC Transmitter

Figure 7.27 shows a traveling wave-based phased-array up-conversion DAC transmitter system. The system combines the LO, mixer, DAC, and PA in a single block, reducing the system power consumption significantly. Multiple large PA loads at very high oscillation frequencies are absorbed into the distributed oscillator tuned tank, eliminating the need for power-hungry buffers. The number of phases, which is eight in this example, can be extended to larger numbers. The power mixer blocks (PMIX) mix the baseband digital data to high frequencies and drive the eight antennas. The power radiated in these antennas is combined in space, yielding a beam-forming phased-array transmitter. The schematic of the unit PMIX-DAC is shown in Figure 7.28. The differential pair mixing power devices Ma and Mb are driven by the traveling wave signal. The binary-sized MOS devices are controlled by the relatively lower speed digital baseband data, effectively modulating the power flowing to the matched antenna load. Controlling the delays to the each antenna in this path, one can steer the transmit beam. Such a multielement phased-array system that can be driven from a single traveling wave signal source results in power-efficient communication at very high frequencies.

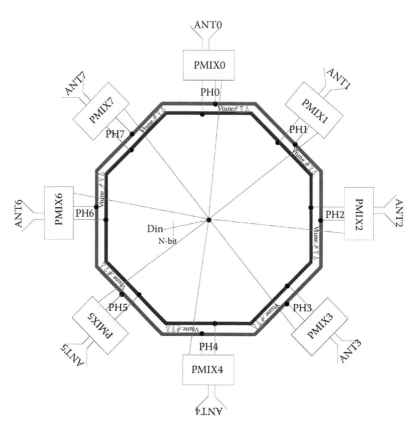

FIGURE 7.27
A traveling wave-based phased-array up-conversion DAC transmitter system.

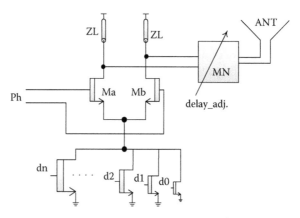

FIGURE 7.28
Unit PMIX-DAC schematic.

7.5 Conclusion

Applications of traveling waves to data converters have been discussed. In the analog-to-digital converters, timing intervals defining the sampling instances can be well defined through use of high-accuracy phases of evenly symmetric traveling wave technologies. The full-swing low-noise clock signals with sharp edges create a unique opportunity for high-speed analog-to-digital converters. Various ADC topologies have been presented, utilizing these multiple phase oscillators. Similarly, well-defined timing intervals are key design parameters for digital-to-analog converters. Using high frequency multiphase full-swing clocks opens a path to high-performance DACs at high frequencies.

Bibliography

1. G. Carreau. Method to reduce error in time interleaved analog-to-digital converters arising due to aperture delay mismatch, US Patent 7843373 B2, Nov. 2010.
2. C. C. Cutler. Transmission systems employing quantization, US Patent 2927962 A, March 1960.
3. F. Eynde. Voltage controlled oscillator (VCO) based on analog-digital converter, US Patent 8760333 B2, June 2014.
4. W. M. Goodall. Telephony by pulse code modulation. *Bell System Technical Journal*, 26:395–409, July 1947.
5. S. Huang. Quantization circuit having VCO-based quantizer compensated in phase domain and related quantization method and continuous-time delta-sigma analog-to-digital converter. US Patent 8,471,743, June 2013.
6. D. R. Main. Multi-phase filter/DAC, US Patent 5,521,946, May 1996.
7. J. C. Schelleng. Code modulation communication system, US Patent 2453461 A, Nov. 1948.
8. B. D. Smith. An unusual electronic analog-digital conversion method. *IRE Transactions on Instrumentation*, PGI-5:155–160, June 1956.

8

Traveling Wave High-Speed Serial Link Design for Fiber and Backplane

High-speed fiber optic and backplane wireline infrastructures represent a significant segment in today's communication systems. In addition to enabling high-speed Internet traffic, these networks handle base station backhauls, enabling much of the wireless traffic as well. Because of the high cost associated with fiber infrastructure, existing systems target increasing the serial data throughput per lane by squeezing more bandwidth into the existing data channels. Hence, the performance requirement from the enabling SERDES chipsets becomes more and more stringent. The existing data rates of 10 and 25 Gb/s per lane are soon going to be replaced by chipsets with 40 and 100 Gb/s per channel transfer rates. Even 400 Gb/s per channel system specifications are being finalized through various standardization bodies.

In the field of high-speed serial links, fiber optic links lead the way despite the challenges at high frequencies, such as intermodal dispersion, chromatic dispersion, or polarization-dependent dispersion along fiber. There are two types of fibers used in today's systems: multimode (MM) and single mode (SM). The difference is set by the physical dimensions of the fiber; having larger-core-diameter (50 to 165 µm) MM fiber allows multiple propagation paths, whereas a relatively narrow fiber core (8 to 10 µm) in SM fibers allows a single propagation path. MM fibers are easier to interface and at lower cost and can deliver higher optical power. However, a strong intermodal dispersion characteristic makes them suitable for only relatively short reach (SR) applications up to around 2 km. High-speed serial link ICs can also be categorized into mux/demux or repeater in terms of functionality. Mux chipsets serialize low-speed parallel data into high-speed single lane, whereas demux chipsets do the opposite, parallelizing the high-speed serial data for network or application processors. Repeater chipsets, on the other hand, are used for signal conditioning and jitter cleaning purposes, enabling long-range data communication. Some repeater chipsets only include linear gain without clock recovery, and hence serve only as an eye opener.

Figure 8.1 shows an example of a long-reach (LR) repeater interface block diagram. In this system the 4 × 25 Gb/s wavelength division multiplexed data arriving at the receiver optical subassembly (ROSA) are first optically demultiplexed into four single 25 Gb/s lanes reaching photodiodes. Low-noise, high-dynamic-range linear transimpedance amplifiers first amplify the incoming signal. Sensitivity of the photodiode and the transimpedance

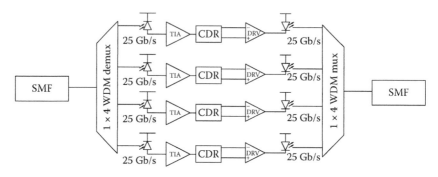

FIGURE 8.1
A 100 Gb/s optical link repeater system.

amplifier together is a critical system parameter impacting the link performance significantly. The signal subsequently goes into a jitter cleaning clock and data recovery (CDR) circuit for retiming with a low-noise local oscillator. Finally, the retimed data head into the laser driver stage, which controls light-emitting diodes (LEDs) or lasers. Although this looks relatively simple in the block diagram, the practical system includes many complex analog/mixed-signal circuits, such as automatic gain control (AGC) loops at the receiver front end, a transmit power-level sensor, and related feedback circuits to circumvent the nonidealities in the system. Some of the commercial products even implement accurate temperature sensors for absolute stability in output laser drive levels, hence best transmitting signal quality as well as a longer lifetime for the emitting device. Although it is possible to find discrete products for each of the mentioned devices in the market, the trend is integration of the pieces to bring power consumption of the overall system down. It is now possible to see products that integrate TIA with CDR or drivers with CDR. Soon, integration of transimpedance amplifiers (TIAs), CDRs, and drivers will be possible. Some even target the integration of front-end optical devices and waveguides on a complementary metal oxide semiconductor (CMOS) die.

A slightly modified version of this diagram, shown in Figure 8.2, provides a full-duplex link to the host network processor. The 25 Gb/s interface between the host interface and retiming module is defined by OIF CEI-28G-VSR, whereas the 4 × 25 Gb/s optical interface is specified in the IEEE 802.3ba 100GE standard.

Figure 8.3 shows a SERDES architecture that interfaces 10 of 10 Gb/s serial data streams at the host processor side, whereas the 25 Gb/s data rate goes into the transmitter optical subassembly (TOSA) and is wavelength division multiplexed into a single fiber lane with an aggregate data rate of 100 Gb/s. The optical interface is again defined by IEEE 802.3ba, whereas the electrical interface at the host side is covered by the CAUI specification. It is even possible to find many field-programmable gate array (FPGA) devices with 10 or 25 Gb/s compatible IOs serving as a network node.

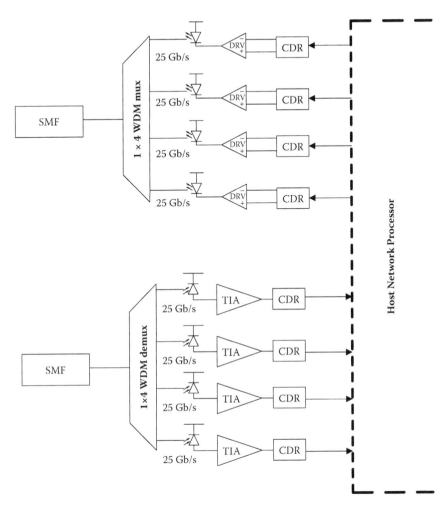

FIGURE 8.2
A 100 Gb/s optical link host interface system.

In the optical interface examples described above, dense wavelength division multiplexing (DWDM) is utilized to reach an aggregate rate of 100 Gb/s. In this case, each of the 25 Gb/s non-return-to-zero (NRZ) data paths implements ON/OFF key (OOK) modulation to a corresponding laser wavelength, and all four modulation products propagate along the same fiber. Although DWDM and introduction of erbium-doped fiber amplifiers (EDFAs) resulted in significant boost in data rate and transmission distance, the spectrum efficiency corresponding to each wavelength remained poor. Moreover, OOK-modulated light along the fiber is susceptible to fiber impairments such as chromatic dispersion (CD) and polarization mode dispersion (PMD) at data rates above 10 Gb/s. This has opened the path to more efficient

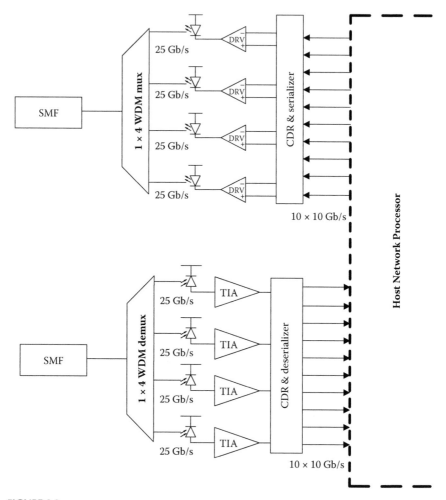

FIGURE 8.3
A 10 × 10 Gb/s electrical to 4 × 25 Gb/s optical link host interface system.

modulation technologies, namely, coherent detection techniques. Coherent techniques that were very commonly being used in wireless and cable systems have recently been applied to optical front ends to increase channel efficiency and link distance. Recent demonstrations of 100 Gs/s coherent links starting in early 2010 have shifted the focus from noncoherent to coherent research for the future 400 Gb/s long-reach applications.

Coherent technologies imply high-order complex amplitude/phase modulation schemes such as differential phase shift keying (DPSK), quadrature phase shift keying, or quadrature amplitude modulation (QAM) in either the electrical or optical domain, or possibly at both ends. Using a Mach-Zehnder modulator, for example, one can introduce binary phase shift keying (BPSK)

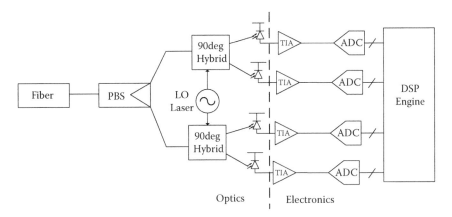

FIGURE 8.4
A coherent system diagram.

at the optical side. In addition to phase and amplitude, the polarization characteristic of the fiber can be utilized to increase the data rate. Fiber can effectively support two orthogonal polarizations. By selectively transmitting
modulated signals using polarization mutilplexed (PM) carriers, one can further double the spectral efficiency of a given modulation scheme. A basic
block diagram of a coherent receiver system is shown in Figure 8.4. The optical input signal from the fiber first goes through a polarization beam splitter
(PBS), which separates the orthogonally polarized signals. Each of these signals then passes through a 90° phase shift hybrid and is mixed down with
an LO laser. Similar to radio systems, the down-converted IQ signal contents are converted to digital through wideband high-speed ADCs. In order
to recover the received bits, carrier phase synchronization is performed in
the digital signal processing (DSP) engine. In transmit systems, high-speed
DACs and drivers would be replacing the ADCs and TIAs, respectively, of
the receiver system shown in the Figure 8.4. Applying DWDM to coherent
systems, overall system transfer rates of Tb/s will commercially be available
in the near future. The current limitations of the coherent technologies are
mostly on the electronics side in design of high-speed ADC and high-speed
DAC.

This chapter introduces some traveling wave circuit techniques that address
power, noise, and speed trade-offs in such SERDES systems. The proposed
techniques benefit from the high-accuracy multiple phases of traveling wave
oscillators discussed in the previous chapters. Hence, any of the TMDWO,
FMDWO, or PDWO techniques can be utilized as a multiphase very high
speed clock source.

This chapter is organized as follows: Section 8.1 goes through a traveling
wave-based multiphase RX and TX front end. Section 8.2 describes a full-rate
RX-TX system with a high-performance phase interpolator. In Section 8.3,

use of traveling wave phases for an ADC-based phase-interleaved front end is described, and Section 8.4 concludes the chapter.

8.1 Traveling Wave-Based Multiphase Rx-Tx Front End

In the literature, many half-rate or even quarter-rate front ends have been introduced utilizing either quadrature LC tank voltage-controlled oscillator (VCO) or a four-stage ring VCO generating the required quadrature or more clock phases [1–4]. These circuits take advantage of running the clocks at a four times slower rate, saving a significant amount of power that otherwise would be required to generate and distribute across various RX and TX circuit blocks. A four-stage differential ring VCO schematic with quadrature phases is shown in Figure 8.5. The idea can further be extended to more phases. However, this may not be sufficient to increase overall speed since the frequency of the ring VCO goes down in reverse proportion to the number of delay stages used.

Traveling wave oscillators, however, decouple the number of oscillation phases from the oscillation frequency, providing multiple symmetric phase tap points along with the traveling wave tracks of a very high frequency distributed oscillator. Not only availability of multiple phases, but also the distributed nature of these oscillators help high-speed multiphase SERDES design. A circuit schematic of an eight-phase RX design is shown in Figure 8.6. The input data and sampling phases are illustrated in the figure. The number of phases, which is eight for this design, can further be extended to larger numbers.

The incoming data are first conditioned through front-end equalization and linear amplification. The multiphase symmetric traveling wave oscillator of choice (TMDWO, FMDWO, or PDWO) is then phase locked to the incoming data through a phase detector and a loop filter. The resulting control voltage

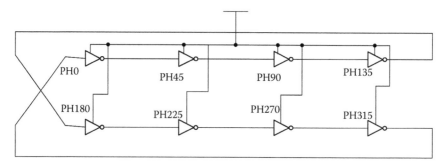

FIGURE 8.5
A four-stage ring VCO as multiphase clock source.

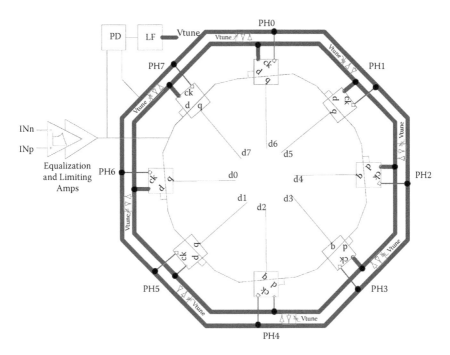

FIGURE 8.6
A dual-loop traveling wave-based multiphase SERDES RX.

tunes the traveling wave speed by use of symmetrically distributed varactors along the transmission line tracks. Once the phase lock is in place, eight best fit phase tap points are used to slice the data into eight lower-speed paths. Depending on the application, these data are either realigned by using additional clock phases or retransmitted through corresponding timing adjustment. These clock phases and input bit stream are shown in Figure 8.7.

The design benefits from two main advantages. First, the relatively low far-out phase noise of the distributed oscillator improves the rms jitter performance of the design significantly. The rail-to-rail signal swing and sharp edges without an additional buffer stage also result in power savings. Second, the effective throughput and speed of these structures can be significantly higher than those of the delay-based multiphase structures. The mentioned eight fine clock phases are also employed in the transmitter to serialize 8-bit parallel data. The schematic of the mentioned TX block with serializer and feed-forward equalizer (FFE) driver is shown in Figure 8.8. The two consecutive phases create the narrow window evaluating the corresponding bit and passing it to the output. In such a configuration, any asymmetry in the clock phases results in systematic cycle-to-cycle jitter in the transmitted data waveform, and hence limits the performance. Cross-coupled PMOS pair devices M1-M2 help well-defined sharp data transitions.

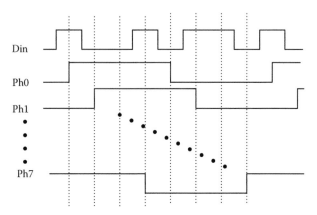

FIGURE 8.7
Data and corresponding traveling wave clock waveforms for an eight-phase wave RX.

8.2 A Full-Rate Phase-Interpolating Topology

Applications with a high jitter tolerance requirement may need fast phase tracking through a digitally controlled phase interpolator (PI). In contrast, a very slow digitally controlled PI loop can help decouple VCO noise dominance from the jitter cleaning action, by locking the oscillator to a clean and stable reference through a wide-bandwidth secondary loop. In any of the cases, an accurate PI is the key block in such systems. Most common phase interpolators utilize four quadrature phases of a high-speed clock to synthesize an optimum sampling clock for the incoming bit stream. A traveling wave-based phase interpolating topology is shown in Figure 8.9. In this system, the traveling wave oscillator is locked to the desired frequency through

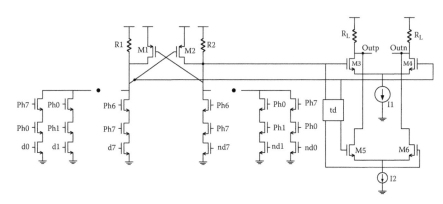

FIGURE 8.8
A multiphase SERDES TX.

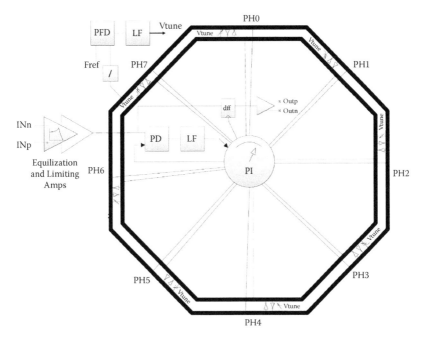

FIGURE 8.9
A full-rate phase-interpolating serial link front end.

an external crystal reference. The frequency drift and jitter from the incoming data are all accounted for by a second phase interpolating loop. A phase detector senses the phase difference between the incoming data and the synthesized clock and applies a correction control word to the phase interpolator following a loop filter (LF). The transient response and bandwidth of this filter can be programmed to fit the desired application. The main feature that differentiates the phase interpolator in this figure is that it uses eight or more very high frequency traveling wave clock phases directly to generate the final desired clock phase. Using more than clock phases results in much better clock waveforms since the interpolation happens between the phases not far apart from each other. The resultant waveforms can have much sharper transition edges, which reduces the noise susceptibility in the following stages. As mentioned previously, the number of clock phases and maximum oscillation frequency are two contradictory constraints in classical multiphase ring oscillators, whereas these are not coupled in the traveling wave oscillators. More phases can be readily available by simply tapping more symmetric points along the traveling wave transmission line tracks.

A circuit implementation of the phase interpolator (PI) in the center of the design is shown in Figure 8.10. All of the differential clock phases from the traveling wave oscillator drive a pair of diff pair devices M1-M2. The mixing cascode devices M3 and M4 connect the desired phases to the output. Only

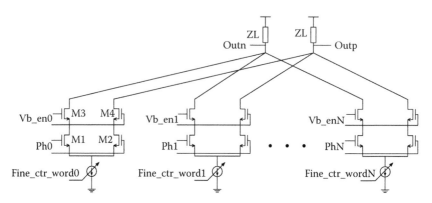

FIGURE 8.10
A current mode phase interpolator circuit.

two pairs of these devices are active at a time, connecting the two consecutive close-by phases to be interpolated. The fine-tuning words controlling the
current sources at the bottom of the cells set the relative weight for each of
the interpolating phases at the final output.

8.3 An ADC-Based DSP Link Front End

Recent optical serial link systems envision moving to higher speeds with the
existing optical networks. This is a significant challenge since the channel loss
and inter-symbol interference (ISI) are both pronounced at higher frequencies.
Classical analog equalization techniques become very hard and power hun-
gry at high speeds. Having a multitap decision feedback equalization (DFE)
with high-speed analog delaying flops is an even harder task. The solution
comes by employing high-speed relatively low resolution ADCs (6 to 8 bits)
in the front end. Such an ADC at full rate is usually prohibitively high power
to incorporate; hence, a phase-interleaved ADC comes as a very natural design choice. Figure 8.11 shows an ADC-based serial link front end using the
symmetric phases of a dual-loop traveling wave oscillator. In order to help the
dynamic range of the ADC, the incoming high-speed serial data may again
be analog preprocessed with continuous-time equalized gain. The signal then
branches into eight ADCs to be sampled and digitized by the corresponding
interleaved phase. The pH0 through pH7 are the eight symmetric differential
clock phases, available at the symmetric tap points along the traveling wave
tracks of the oscillator. The resulting digital information leads into a DSP core,
which handles filtering, channel equalization, and phase lock. The core feeds
back an optimum control signal to the varactors for precise frequency and
phase lock. The details of the high-speed ADC circuit design are presented in
Chapter 7.

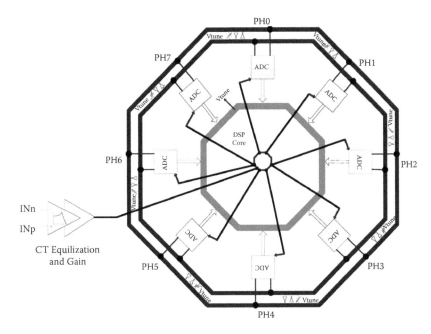

FIGURE 8.11
A link front end with an interleaving ADC-based DSP.

8.4 Conclusion

Applications of traveling waves to SERDES systems have been discussed. As the speed of these systems goes up, the timing requirements become more and more difficult. Considering the rise and fall times for such speeds, carrying the clocks in these chipsets is itself a challenge. The chapter presented systems in which low-noise multiphases of traveling wave oscillators were utilized to reach the same timing resolution with lower power consumption. Similar systems can always be constructed by using multiphase ring oscillators, but with a speed limit and higher noise penalty.

Bibliography

1. C. C. Huang. Multi-phase-locked loop for data recovery, US patent 6442225 B1, Aug. 2002.
2. A. P. Van der Wel and G. W. D. Besten, Multi-phase clock and data recovery system, US patent pending 20120154059 A1, June 2012.

3. J. Lee, and B. Razavi, High-speed clock and data recovery circuit, US Patent 7286625 B2, Oct. 2007.

4. P. Johnson, Z. Chen, and B. Britton, Clock-and-data-recovery system having a multi-phase clock generator for one or more channel circuits, US Patent 7599457 B2, Oct. 2009.

Index

A

Absolute temperature, 41–42
ADCs (analog-to-digital converters),
    4, 99
    based DSP link front end, 130, *131*
    fiber polarization, 125
    high-speed phase interleaving
        topology, 109–115
    pipeline, 105–106, 110–111
    sigma-delta-type, 102–104
    successive approximation register
        (SAR)-type, 99–101
Admittance Smith chart, 20, *21*
Amplifiers
    low-noise, 71–72, 76–79, 93
    transimpedance, 122
Analog-to-digital converters (ADCs),
    4, 99
    based DSP link front end,
        130, *131*
    fiber polarization, 125
    high-speed phase interleaving
        topology, 109–115
    pipeline, 105–106, 110–111
    sigma-delta-type, 102–104
    successive approximation register
        (SAR)-type, 99–101
Antennas
    force mode distributed wave
        oscillator, 59, *62–63*, 63–69
    traveling wave frequency shift
        reflectometer, 89

B

Binary phase shift keying (BPSK), 4,
    124–125
Bluetooth, 2, 3
Boltzmann's constant, 41–42
BPSK (bi-phase shift key), 4, 124–125

C

CD (chromatic dispersion), 123
Chromatic dispersion (CD), 123
Circular standing wave oscillator
    (CSWO), 32, *33*
Clock and data recovery (CDR) circuit,
    122
Clock phases, high-speed phase
    interleaving topology, 109–115
CMOS (complementary metal oxide
    semiconductor) process, 4, 54, 64,
    99
Coherent technologies, 124–125
Complementary metal oxide
    semiconductor (CMOS) process, 4,
    54, 64, 99
Coplanar strip line, 22, *25*
Coplanar waveguide, 22, *25*
Cross-coupled pair noise, 43–46
CSWO (circular standing wave
    oscillator), 32, *33*

D

DACs (digital-to-analog) converters, 4,
    99
    feedback, 108–109
    successive approximation register
        topologies and, 101
    traveling wave multiphase, 115–117
    traveling wave phased-array
        transmitter, 117, *118*
DDBCs (differential drive branch
    couplers), 88
Decision feedback equalization (DFE),
    130
Delay-based FMDWO circuit, 58–59
Dense wavelength division multiplexing
    (DWDM), 123
Differential coplanar waveguide, 22, *25*

Differential drive branch couplers (DDBCs), 88
Differential wave oscillator (DWO), 51–53
Digital-to-analog converters (DACs), 4, 99
  feedback, 108–109
  successive approximation register topologies and, 101
  traveling wave multiphase, 115–117
  traveling wave phased-array transmitter, 117, *118*
Dispersionless transmission line, 9–11
Distributed oscillator structure using transmission lines, 23, *25*, 26
DWDM (dense wavelength division multiplexing), 123
DWO (differential wave oscillator), 51–53

E

EDGE (Enhanced Data for GSM Evolution), 3
802.11 standard, 2
Electromagnetic spectrum, 4–5
Enhanced Data for GSM Evolution (EDGE), 3
Ethernet/optical networks, 4

F

Federal Communications Commission (FCC), 1
Feedback DACs, 108–109
Feed-forward equalizer (FEE), 127
FEPS (front-end phase shift) system, 74
Fiber optic links, 121, 123
Field-programmable gate array (FPGA), 122
FLL (frequency lock loop), 107–109
FMDWA antenna, 59, *62–63*, 63–64, 71
FMDWO (force mode distributed wave oscillator), 57
  mechanisms, 57–59
  single-ended force mode structures, 59–68
  traveling wave phased-array transceiver, 74–79

Force inverting amplifiers, 58–59
Force mode distributed wave oscillator (FMDWO), 57, 71
  mechanisms, 57–59
  single-ended force mode structures, 59–68
  traveling wave phased-array transceiver, 74–79
Forward waves, dispersionless transmission line, 10
4G mobile telecommunication standards, 3
Four-stub supply distribution scheme, 73–74
FPGA (field-programmable gate array), 122
Frequency-dependent negative resistance (FDNR)-based noise-shaped filter, 72
Frequency lock loop (FLL), 107–109
Frequency multiplication techniques, 84–87, *88*
Front-end phase shift (FEPS) system, 74

G

General Packet Radio Service (GPRS), 3
Global Positioning System (GPS), 3
GPRS (General Packet Radio Service), 3
GPS (Global Positioning System), 3

H

Harmonics, 86–87
Heterodyne traveling wave reflectometer (HTWR), 92–93
High Speed Package Access (HSPA), 3
High-speed phase interleaving topology, 109–115
HSPA (high Speed Package Access), 3
HTWR (heterodyne traveling wave reflectometer), 92–93

I

Impedance, 8–9, 11, 13–14
  phase noise and, 39–41, 47–50
  Smith chart, 16–20, *21*
Inductively pumped distributed wave oscillator (IPDWO), 74

Infinite transmission line, 7–9
Institute of Electrical and Electronics
    Engineers (IEEE), 2
IP-based mobile broadband, 3
IPDWO (inductively pumped
    distributed wave oscillator), 74
ISM band, 3–4

K

Kirchhoff's current laws (KCL), 8, 47
Kirchhoff's voltage laws (KVL), 8

L

LANs (local area networks), 1
Lasers, 122
Least significant bits (LSBs), 105–106
Light emitting diodes (LEDs), 122
LNA (low-noise amplifier), 71–72,
    76–79, 93
Local area networks (LANs), 1
Long-reach (LR) repeater interface block
    diagram, 121, *122*
Lossless transmission line, 11–13
    voltage waveform, 14–15
Low-noise amplifier (LNA), 71–72,
    76–79, 93
Lumped circuit models, 7
Lumped LC tank oscillator, *25*

M

Metal oxide semiconductor field effect
    transistor (MOSFET), 22
Microstrip line, 22, *24*
MIMO (multiple-input multiple-ouput)
    technology, 2
Moore's law, 4
MOS device, 93–95
MOSFET (metal oxide semiconductor
    field effect transistor), 22
Most significant bit (MSB), 101, 104–105
Multimode (MM) fibers, 121
Multiple-input multiple-output (MIMO)
    technology, 2

N

NMOS (non-complementary metal oxide
    semiconductor), 72–74
    differential pair, 112–113

Noise transfer function (NTF), 104–105
Non-complementary metal oxide
    semiconductor (NMOS), 72–74
NTF (noise transfer function), 104–105

O

ON/OFF key (OOK) modulation, 123
Oscillators, 22–26
    commonly used wave oscillator
        topologies, 26–27
    coplanar strip line transmission line
        and, 23, *25*
    coplanar waveguide transmission line
        and, 23, *25*
    differential coplanar waveguide
        transmission line and, 23, *25*
    differential wave (DWO), 51–53
    experimental results, 53–55
    high-speed phase interleaving
        topology, 109–115
    lumped LC tank transmission line
        and, 23–25
    microstrip line transmission line and,
        23, *24*
    phase noise in traveling wave, 38–53
    rotary traveling wave oscillator
        (RTWO), 27–31, 46–51
    standing wave oscillator (SWO),
        31–32, *33*, 38–46
    traveling wave-based multiphase
        Rx-Tx front end, 126–127, *128*
    traveling wave noise-shaping
        modulator, 107–109
    trigger mode distributed wave
        oscillator (TMDWO), 26, 34–38,
        54–55
    voltage-controlled, 34–38

P

P code, 4
Phased-array DAC transmitter, 117, *118*
Phase-interpolating (PI) topology,
    128–130
Phase-locked loop (PLL), 57
Phase noise, 38–53
    cross-coupled pair noise, 43–46
    differential wave oscillator, 51–53
    RTWO, 46–51

SWO, 38–46
tail transistor noise, 42–43
transmission line, 41–42
Pipeline ADCs, 105–106, 110–111
PI (phase interpolating) topology,
128–130
PLL (phase-locked loop), 57
PMOS differential pair, 112–113
Power transfer efficiency, 13
Printed circuit boards (PCB), 59, *60*
Propagation constant, infinite
transmission line, 9
Pseudorandom number (PRN) sequence,
4

Q

Quadrature LC tank VCO, 34–36
Quality factor, resonator, 29–31, 41
Quarter-pumped distributed wave
oscillator (QPDWO), 71, 72–74
four-stub supply scheme, 73–74
traveling wave phased-array
transceiver, 74–79, *80–81*

R

Radio frequency (RF) transmitters, 57,
71, 72
Reflectometers, traveling wave, 88–95
Resonant-pumped distributed wave
oscillator (RPDWO), 74
Resonator quality factor, 29–31, 41
RF (radio frequency) transmitters, 57, 71,
72
Rotary traveling wave oscillator
(RTWO), 27–31
phase noise, 46–51
RPDWO (resonant-pumped distributed
wave oscillator), 74
RTWO (rotary traveling wave oscillator),
27–31
phase noise, 46–51

S

Sensing method, wafer-level THz, 95–96
SERDES architecture, 121, 122, 125
Shorted transmission line, 13–14
Short Message Service (SMS), 3
Sigma-delta-type ADCs, 102–104

Signal transfer function (STF),
104–105
Single-ended force mode structures,
59–68
Single mode (SM) fibers, 121
Smith chart, 16–20, *21*
SMS (short message service), 3
Standing wave oscillator (SWO), 31–32,
33
phase noise in, 38–46
STF (signal transfer function), 104–105
STWFM (switching traveling wave
frequency multiplier), 90–92
Successive approximation register
(SAR)-type ADCs, 99–101
Supply stubs, 73–74
Switching traveling wave frequency
multiplier (STWFM), 90–92
SWO (standing wave oscillator), 31–32,
33
phase noise in, 38–46
Symmetry-based FMDWO circuit, 59, *60*

T

Tail transistor noise, 42–43
Temperature, absolute, 41–42
3G mobile telecommunication standards,
3
THz-frequency, 83–84
frequency multiplication techniques,
84–87, *88*
gap, 5
traveling wave reflectometers,
88–95
wafer-level sensing method, 95–96
TMDWO (trigger mode distributed
wave oscillator), 26, 71
experimental results, 54–55
topology, 34–38
traveling wave phased-array
transceiver, 74–79
TOSA (transmitter optical subassembly),
122
Transimpedance amplifiers (TIAs), 122
Transmission lines
coplanar strip line, 23, *25*
coplanar waveguide, 23, *25*
differential coplanar waveguide,
23, *25*

dispersionless, 9–11
distributed oscillator structure using, 23, *25*, 26
infinite, 7–9
lossless, 11–13
lumped LC tank, 23–25
microstrip line, 23, *24*
noise, 41–42
shorted, 13–14
Smith chart, 16–20, *21*
structures, 22
voltage standing wave ratio, 14–15
Transmitter optical subassembly (TOSA), 122
Traveling wave-based multiphase Rx-Tx front end, 126–127, *128*
Traveling wave frequency multiplier (TWFM), 87, *88*
    traveling wave reflectometers, 88–95
Traveling wave frequency shift reflectometer (TWFSR), 89
Traveling wave multiphase DAC, 115–117
Traveling wave noise-shaping modulator (TWNSM), 107–109
Traveling wave phased-array DAC transmitter, 117, *118*
Traveling wave phased-array transceiver, 74–79, *80–81*
Traveling wave reflectometers, 88–95
Trigger mode distributed wave oscillator (TMDWO), 26, 71
    experimental results, 54–55
    topology, 34–38
    traveling wave phased-array transceiver, 74–79
TWFM (traveling wave frequency multiplier), 87, *88*
    traveling wave reflectometers, 88–95

TWFSR (traveling wave frequency shift reflectometer), 89
TWNSM (traveling wave noise-shaping modulator), 107–109

U

UHFs (ultra-high frequencies), 1
Ultra-high frequencies (UHFs), 1
U.S. Nuclear Detonation Detection System (USNDS), 4
USNDS (U.S. Nuclear Detonation Detection System), 4

V

VCO (voltage-controlled oscillator), 34–38, 126
Voltage-controlled oscillator (VCO), 34–38, 126
Voltage standing wave ratio (VSWR), 14–15
VSWR (voltage standing wave ratio), 14–15

W

Wafer-level THz sensing method, 95–96
WCDMA (wideband Code Division Multiple Access), 3
Wideband Code Division Multiple Access (WCDMA), 3
Wi-Fi LAN (wireless local area networks), 1, 2–4
Wilkinson coupler, 93
WiMAX, 3
Wireless local area networks (Wi-Fi LAN), 3

Z

Zero efficiency, 13

T - #0859 - 101024 - C13 - 234/156/7 - PB - 9781138893702 - Gloss Lamination